AF806678

CSR, Sustainability, Ethics & Governance

Series editors
Samuel O. Idowu, London Metropolitan University, London, United Kingdom
René Schmidpeter, Cologne Business School, Germany

More information about this series at http://www.springer.com/series/11565

Nicolas Josef Stahlhofer • Christian Schmidkonz •
Patricia Kraft

Conscious Business in Germany

Assessing the Current Situation and Creating an Outlook for a New Paradigm

Springer

Nicolas Josef Stahlhofer
NTT DATA
Munich, Germany

Christian Schmidkonz
Munich Business School
Munich, Germany

Patricia Kraft
Munich Business School
Munich, Germany

ISSN 2196-7075 ISSN 2196-7083 (electronic)
CSR, Sustainability, Ethics & Governance
ISBN 978-3-319-69738-3 ISBN 978-3-319-69739-0 (eBook)
https://doi.org/10.1007/978-3-319-69739-0

Library of Congress Control Number: 2017958031

© Springer International Publishing AG 2018
This work is subject to copyright. All rights are reserved by the Publisher, whether the whole or part of the material is concerned, specifically the rights of translation, reprinting, reuse of illustrations, recitation, broadcasting, reproduction on microfilms or in any other physical way, and transmission or information storage and retrieval, electronic adaptation, computer software, or by similar or dissimilar methodology now known or hereafter developed.
The use of general descriptive names, registered names, trademarks, service marks, etc. in this publication does not imply, even in the absence of a specific statement, that such names are exempt from the relevant protective laws and regulations and therefore free for general use.
The publisher, the authors and the editors are safe to assume that the advice and information in this book are believed to be true and accurate at the date of publication. Neither the publisher nor the authors or the editors give a warranty, express or implied, with respect to the material contained herein or for any errors or omissions that may have been made. The publisher remains neutral with regard to jurisdictional claims in published maps and institutional affiliations.

Printed on acid-free paper

This Springer imprint is published by Springer Nature
The registered company is Springer International Publishing AG
The registered company address is: Gewerbestrasse 11, 6330 Cham, Switzerland

To Josef Stahlhofer

Foreword

A world which is constantly shaped by volatility, uncertainty, complexity and ambiguity asks for wise and responsible leadership. Navigating through this field of tension is anything but easy. Traditional management and leadership models such as linear forecast models increasingly show their weaknesses. Nevertheless, there are also numerous examples of leaders and companies who manage to cope with this environment based on their systemic intelligence, structural agility and their pursuit of a higher purpose. These organizations are led by purpose rather than fear which gives them a competitive advantage in terms of actively transforming their structures to external developments. The foundation of their success is based on a core principle—they are conscious of themselves and others and act upon it.

When reading the news regarding impactful businesses, many people might come to the conclusion that data-driven and platform-based businesses from the USA and Asia will leave no room for other players in the future. However, Germany is home to hundreds of companies who have been achieving financial success while actively fostering positive change for their stakeholders at the same time.

"Conscious business in Germany" portrays four of these examples in a comprehensive way. The companies portrayed are united by a shared belief that their business model contributes to the elevation of humanity. These organizations have managed to build a conscious way of leadership as well as an inclusive stakeholder model around their business purpose. Their corporate cultures are dominated by a high level of employee inclusion regarding joint decision-making, transparency, trust, flat hierarchies and authentic communication. Establishing such companies does not happen overnight and requires constant support by the management. Hence, the authors discovered each company by conducting interviews with the founders, CEOs and top managers of these four Germany conscious businesses to gain valuable insights of their leadership and cultural approach.

Conscious businesses excel in involving their employees to participate in the companies' purpose and dynamic culture. This culture is formed by a humanistic conception of man, an open platform for innovative ideas, personal responsibility and flat hierarchies. Moreover, passion and happiness are seen as essential factors for workplace performance and employee well-being rather than a superficial catch phrase in company presentations. Cultures are formed by interactions, attitudes and beliefs, and conscious cultures offer enough room for people to discover the corporate soul while growing personally and professionally.

Understanding customer needs and the underlying motivations and values is a key task for every business. Translating this information into long-lasting customer satisfaction, retention and sustainable sales in combination with a humanistic approach might seem contradictory when applying a traditional logic of managing business. These companies have been proving a different reality by switching their minds from a trade-off mentality to a synergetic mentality of matching stakeholder needs and putting their customers at the heart of their operations. In order to provide a comprehensive foundation for this aspect, the authors integrate qualitative and quantitative information and thus create a line of argument.

The portrayed companies use their purpose and humanistic core to guide business decisions and thus ensure coherence and authenticity among internal and external stakeholders. The barriers of hiding societal-unaccepted actions are constantly rising higher due to a growing demand for transparency and responsibility.

The business case for conscious businesses as outlined in this book has implications on different dimensions. For example, recruiting generation y and z talents is easier as these generations are highly attracted by companies that combine purpose and meaningfulness with a successful business model. In addition, keeping a motivated workforce is strongly influenced by a supportive culture and positive work atmosphere that is a core aspect of conscious businesses. With regard to customer acceptance and loyalty, brands that prove to combine sustainable products and services with customer centricity and quality show higher rates of customer loyalty and retention.

EY works on building a better working world by helping companies to embrace change and transformation according to internal and market needs. Hence, we see purpose and related aspects as an essential part of our portfolio which is constantly growing in importance. In addition, the EY Beacon Institute was founded as a community of business leaders, board members and academics to provide research, insights and advice to inspire and amplify the growing movement of purpose-led businesses. In line with this, EY and I am personally convinced that there is an increasing need for the described business practices. Hence, I am very pleased that this book both covers the foundations and outlines how these principles are successfully lived by German companies.

The authors succeeded in offering a comprehensive book for everyone who is interested in discovering the world of purpose-led companies and educating oneself concerning the key drivers, the needed mindset and specific corporate actions towards a new paradigm in business.

Chairman of the Management Board Hubert Barth
EY Germany, Munich, Germany
October 2017

Preface

There is no day in our life which passes without interacting with numerous products and services that are offered to us by different businesses: when it's dark, we switch on the light and use the energy from a provider without thinking about the means of production. The soap and toothpaste we use have not been created by us but bought in one of the many drug stores nearby. We purchase our clothes with a more or less conscious mindset from a well-known brand which had to set up various supply chains to offer them. As all of these daily items cost money, we have "our" bank which enables us to buy things in various ways. Nevertheless, the bank will only provide us with monetary means that we have earned through our daily work at the company we are employed at. At the end of the day, companies are actively involved in almost everything we do.

There is a constant potential for annoyance and pleasure which is only possible through this active engagement. One of the most negative corporate incidents in Germany during the past years has been the so-called dieselgate. Technical manipulations have been consciously tolerated and partly forced by a large number of automotive executives, engineers, and others. Not only have customers been betrayed but also the habitants of countries where the cars were sold in, due to higher than expected pollution. The German automotive industry would be a loss-making business if all harmful ecological and economic impacts would be priced in.

On the contrary, such scandals empower a constantly increasing number of organizations and individuals to do business in a more conscious manner. Doing business is a necessary prerequisite for societal well-being and better living standards for billions of people. Yet, the future flourishing of mankind requires a more conscious approach to business. Implementing this holistic idea of consciousness does not only have an impact on customers, the environment, employees, suppliers, investors, partners, competitors, media, and institutions but also on the society at large. Mankind is put at the center of Conscious Businesses. Monetary success can only be created through the collective well-being of the community in which the business is operating and in turn actively contributing with its activities.

A CEO who is only concerned with key performance indicators such as revenue and profit at the end of a quarter is the embodiment of a CEO of the past century. It is not a great achievement to act upon the "whatever it takes" principle to create profit for a business. Looking at the mere profit does not reflect negative or destructive aspects which were caused on the way to create it. A company which is led in such a way will not be able to persist in the long term and will vanish without being missed.

A CEO of a Conscious Business is concerned with corporate culture and the match between performed activities and corporate values at the end of a quarter. In addition, he or she focuses on how these values are lived by employees and supported by external stakeholders. From this perspective, customers represent a part of any Conscious Business as they help to cocreate something that is bigger than them. Customers vote with their wallets and thus have the ability to actively shape the direction of companies. The businesses that are portrayed in this book embrace this mindset and offer their customers multiple opportunities to become a valuable part of their purpose.

Four German companies from different industries were selected for this book to present their contribution to the "Conscious Business Movement": the energy provider Polarstern, the outdoor equipment producer VAUDE, the drug store chain dm drogerie markt, and the GLS Bank embody representatives of the Conscious Business movement in Germany. The movement's origin lies in the United States. Founders such as John Mackey (Whole Foods Market), Yvon Chouinard (Patagonia), or Herb Kelleher (Southwest Airlines) are regarded as charismatic and inspiring personalities who created more than just monetary value with their businesses. All of them have one thing in common. They broke a vast number of business administration principles, put mankind and the environment at the center of their actions, and have been financially successful at the same time. Whole Foods Market was acquired by Amazon in June 2017 for $13.4 billion, Southwest Airlines has never experienced a loss by the end of the fiscal year since their foundation in 1971, and Yvon Chouinard has become a billionaire even though he perceives the further cultivation of his corporate philosophy as way more important.

The future belongs to companies that intrinsically apply the principles of Conscious Business which are outlined in this book. "Corporate Social Responsibility" (CSR) has gained widespread acceptance in the 1990s. CSR promotes the integration of social and ecological matters besides the pursuit of financial success and can be seen as a remnant of the past century. The upcoming generations demand for more than just CSR in light of the variety and complexity of global challenges. They demand for conscious and reflected behavior which considers the effects of each action toward the well-being of mankind, the environment, and the contribution to a meaningful life. In addition, self-actualization and the return to values which are in line with the nature and lived by millions of people worldwide represent essential drivers in times of increasing digitalization.

"Conscious Business in Germany" shows that there are already companies in place that are led by founders and CEOs who act upon the "purpose before profit" paradigm. These businesses serve as a contemporary proof and inspiration for a future-oriented way of doing business. This new paradigm puts human beings and money in the right relationship, as part of a highly complex, dynamic, and mutually dependent ecosystem where money is perceived as a means to an end and not as an end in itself. The value which is created by conscious businesses goes far beyond established metrics for success as it pursues a higher purpose in an inherent manner.

Thinking and acting as a conscious business might require more effort than being part of a traditional business. Nevertheless, the resulting meaningfulness of economic activities is of invaluable importance for societal well-being and the future of business. Be inspired by this book and whenever you nod your head while reading it, keep in mind that you are not alone.

August 2017

Nicolas Josef Stahlhofer
Christian Schmidkonz
Patricia Kraft

Contents

List of Abbreviations

CAGR	Compound Annual Growth Rate
CSR	Corporate Social Responsibility
EESC	European Economic and Social Committee
EOG	European Outdoor Group
GABV	Global Alliance for Banking on Values
P/E	Price Earnings Ratio
TACTILE	Trust, Accountability, Caring, Transparency, Integrity, Loyalty, Egalitarianism

Chapter 1
Introduction

We are in the midst of a historic transition where it is becoming clear that old paradigms no longer work as well as before and people's minds are open to new possibilities. Global challenges and evolving opportunities of our era demand visionary thoughts and bold action (Mackey and Sisodia 2014). The relating shifts will continuously reframe the face of the global economy towards a data-driven and interconnected business environment that cannot be stopped by past paradigms and ego-centristic worldviews. The emergence of social networks of people and things will increasingly transform the fundament of communication and will enable a new level of global collaboration. In addition, a far-reaching shift in consciousness can be observed which becomes visible by a confrontation between destructive and creational economic forces (Scharmer 2016).

Financial crises, company scandals, corruption, and unethical behavior triggered a widespread loss of trust in the free-market system, the integrity of business leaders and their competence in decision-making (Benett et al. 2009). Although global crises and company scandals have been existent for decades, the dynamics of modern information society are significantly raising consciousness among people (Nefiodow and Nefiodow 2014). More and more it seems as though the well-being of civilization depends on the cultures and actions of businesses (Sisodia et al. 2014). A present which is characterized by accelerating globalization, digitalization, vanishing boundaries, rapid changes in business and society, uncertainty, volatility and an increasing level of complexity and ambiguity demand conscious businesses which offer clarity, stability, and mindfulness regarding a common good. Conscious corporations offer sense amid confusion, foster community, inspire and provide role models for its stakeholders (Benett et al. 2009). Traditional business models and self-centricity do not reflect an adequate strategy to cope with challenges such as global warming, which are already present and will likely increase in the future, if organizations continue to do business as usual (Matthies 2016).

Volkswagen's diesel scandal has demonstrated that unethical behavior results in a significant loss of credibility and also impacted the trust in German companies in

© Springer International Publishing AG 2018

N.J. Stahlhofer et al., *Conscious Business in Germany*, CSR, Sustainability, Ethics & Governance, https://doi.org/10.1007/978-3-319-69739-0_1

a negative way. The company's crisis did not only emphasize the damage of institutional deception but has been a representative of dysfunctional leadership communication and behavior. It has become clear that fear-driven organizations do not longer embody the zeitgeist of purpose-driven individuals and societies.

People increasingly demand a pursuit of purpose beyond profits and search for companies that operate in a meaningful and mindful way and contribute to the creation of a common good (Sisodia et al. 2014). Being able to express the full spectrum of individual capabilities constitutes a key determinant for employees worldwide and challenges existing leadership paradigms towards a conscious approach of leading and conducting business (Dethmer et al. 2015). The concept of conscious business first evolved under this term in the United States.

Companies such as Whole Foods Market or Southwest Airlines have been demonstrating that conscious businesses are able to fulfill the needs of people and are successful at the same time. This book aims at identifying German companies that reflect the principles of conscious business. The explorative approach of this study translates into an evaluation of four German companies and an overview of potential conscious businesses in order to assess the current state and contribute to the creation of a future outlook.

1.1 Scope and Objectives of the Book

The scope of this book mainly focuses on the current state of German companies concerning the application of conscious business principles. With regard to this, it is subject to question which drivers formed the concept in general, and which aspects can be linked to developments in Germany in particular. As conscious business embodies a different logic of business compared to traditional models, it is questionable which long-term benefits arise from implementing it. In addition, it has to be clarified how American companies successfully implemented these practices, and whether these organizational changes can also be identified in four selected German companies. Core characteristics of conscious business are also reflected in different concepts such as the B Corporations and the Economy for the Common Good. Hence, it is subject to question how these concepts differ from conscious business, and to what extent they are represented by German companies. Finally, it is undefined which role conscious business could play in the future in general and in Germany in particular. Based on the scope of the problem, the objectives of the book are divided in several areas of research. The theoretical foundation of conscious business embodies the groundwork for answering the mentioned questions above. This foundation aims at explaining the surrounding economic environment, driving factors behind the concept, its consequences for businesses, implemented practices, and indicators for its future development. The deductive approach of the book translates into a combined analysis of quantitative and qualitative data by integrating various studies, guided interviews, and relevant company information. Consequently, this approach represents the basis for implications and further discussion.

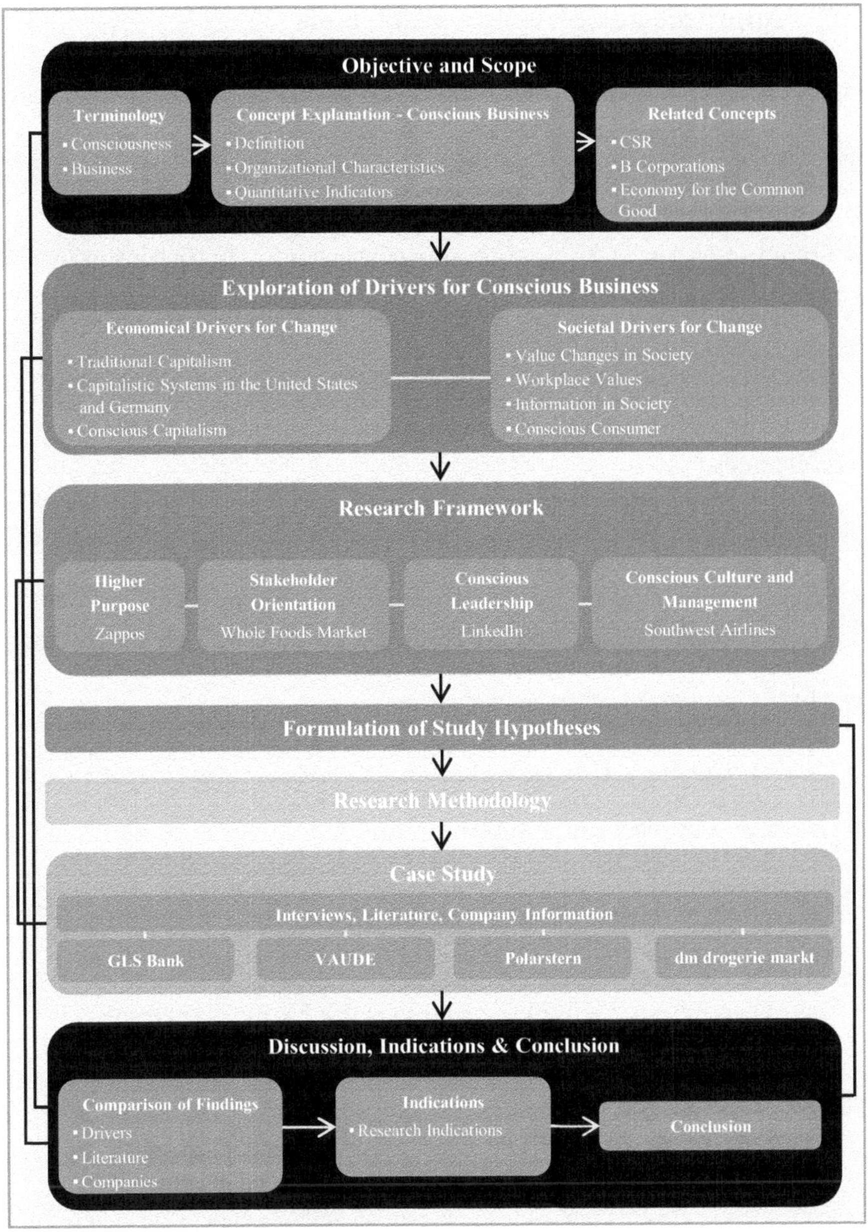

Fig. 1.1 Structure of the book (author's diagram)

1.2 Approach and Structure of the Book

The approach of this book combines an explanation of the objective and scope with a link to findings from literature and related studies. This represents a necessary prerequisite in order to formulate the main aspects of the theoretical framework which is applied in the case studies. The findings from the case studies are linked to the results from the theoretical framework to highlight relations, similarities, and differences which build the foundation for the conclusion. The book is structured as outlined in Fig. 1.1. The arrows symbolize the chronological order of each chapter. The straight lines are used to highlight the connections between the respective elements and their relation.

References

Benett A, Gobhai C, O'Reilly A, Welch G (2009) Good for business: the rise of the conscious corporation. Palgrave Macmillan, New York

Dethmer J, Chapman D, Warner Klemp K (2015) The 15 commitments of conscious leadership: a new paradigm for sustainable success

Mackey J, Sisodia R (2014) Conscious capitalism: liberating the heroic spirit of business. Harvard Business School Publishing, Boston, MA

Matthies P (2016) Personal interview

Nefiodow L, Nefiodow S (2014) The sixth kondratieff: a new long wave in the global economy. Rhein-Sieg, Sankt-Augustin

Scharmer O (2016) Theory U: leading from the future as it emerges. Berrett-Koehler, Oakland

Sisodia R, Sheth J, Wolfe D (2014) Firms of endearment: how world-class companies profit from passion and purpose. Pearson, New York

Chapter 2
Putting Conscious Business Into Context

2.1 Consciousness

This paragraph outlines different perspectives on the meaning of the term *consciousness* with regard to its application in business. Velmans (2009) points out that definitions of consciousness need to be sufficiently broad to include all examples of conscious states and sufficiently narrow to exclude entities, events and processes that are not considered conscious. "To be conscious means to be fully awake and mindful, to see reality mire clearly, and to more fully understand all consequences—short-term and long-term—of our actions" (Mackey and Sisodia 2014, p. 29). Although the development of consciousness is linked to changes and preconditions in the past, it is a state of awareness, which is only taking place in the present moment (Senge et al. 2004). This state reflects a greater awareness of our inner self, our external reality, and the impacts we have on the world. In addition, it translates to a "greater commitment regarding the truth and to acting more responsibly according to what we know to be true" (Mackey and Sisodia 2014). Richard Barrett (2017a) views this concept from a motivational perspective, as he describes it as "awareness with a purpose which arises in living beings and is driven by meeting individual needs at a certain stage of psychological development".

This awareness acts as a moderating entity regarding opportunities and obstacles which influence the successful achievement of needs. Siegel (2010) adds that incorporating a high level of consciousness regulates emotional processes and cultivates a more flexible and stable mind. With reference to the individual environment, Branden (1999, p. 11) offers a behavioral focus, as living and acting consciously describes an "active mental state which allows individuals to be aware of interests, actions, values, purposes and goals". Moreover, this state is focused on expanding the levels of awareness and understanding of internal and external environments. Kofman (2006, p. 3) supports a similar definition, as "living consciously means to be open to perceiving the world around and within us, to understand our circumstances, and to decide how to respond to them in ways, that

© Springer International Publishing AG 2018

N.J. Stahlhofer et al., *Conscious Business in Germany*, CSR, Sustainability, Ethics & Governance, https://doi.org/10.1007/978-3-319-69739-0_2

honor our needs, values, and goals." Thus, human consciousness offers a capacity for abstraction through which a person can transcend experiences by using the full intellectual ability to understand, judge, and reason. This ability is required to organize and manage complex situations in intrapersonal and interpersonal situations. Beyond choosing what to do, the conscious pursuit of happiness entails choosing how to do what one does (Kofman 2006). Human beings are existentially dependent on social interactions and cooperation. This principle also accounts for business environments and therefore needs an additional interpersonal perspective. With regard to a spiritual perspective, Secretan (2001, p. 19) states that highly-conscious individuals are able to "realize the mystical and ineffable aspects of our existence as part of a magical flowing whole." Koch (2012, p. 31) notes with regard to this, that "consciousness is filled with meaningful percepts and memories of sometimes unbearable intensity." The definition of Emile Durkheim transfers an individual level towards a collective dimension. Accordingly, a collective sense of consciousness refers to shared perspectives, beliefs, values, norms, moral attitudes which act as a unifying force within society (Smith 2014). Developing individual consciousness also transcends to a strong sense of interconnectedness regarding living beings, reduces feelings of isolation and alienation and thus serves a collective well-being, compassion, and altruism (Vieten et al. 2006).

2.2 Business

Due to the nature of interrelatedness between business and society, the direction and definition of businesses has developed and changed in the last decade. Businesses experience the needs of customers as a central driver of their behavior. Hence, norms, values, and beliefs of customers and of society as a whole have been shaping the business environment around the globe (Trehan and Trehan 2009). Wheeler (1962, p. 26) offers a rather traditional definition as he views "business as an institution organized and operated to provide goods and services to the society under the incentive of private gain". Goyal and Goyal (2009, p. 4) add an interactional dimension by stating that business is essentially a "continuous human economic activity." Concerning the motivation of this activity, Trehan and Trehan (2009) note that focusing on sole profit maximization cannot be considered the main objectives of business anymore, as society as well as other stakeholders demand a growing sense of responsibility and alignment of social values with business practices. "Profit is not the explanation, cause or rationale of business behavior and business decisions, but rather the test of their validity" (Drucker 2008a, p. 18). In addition, it is important to notice that companies are extensions of people and therefore should be assessed and treated as such. Applying a dehumanized perspective is therefore not portraying a business in a conscious and concise way (Sisodia et al. 2014).

According to Mackey and Sisodia (2014, p. 45), business transactions are based on voluntary exchange. Enabling a "voluntary exchange for mutual benefit creates the ethical foundation of business, and that is why business is ultimately justified to rightfully exist within a society". In summary, the nature of business can be characterized as an economic activity, which is carried out by individuals or a group of individuals, who voluntarily exchange products or services for a defined value at a given risk. The activity itself aims at establishing mutual needs of the transaction partners as well as their impacted stakeholders (Gupta 2007).

2.3 Conscious Business

Businesses have an enormous potential to do good in the world. Many practices and approaches that lead to a greater societal good are performed *unconsciously* by creating valuable solutions for multiple stakeholders (Sisodia et al. 2014). Internalizing a set of humanized values is increasingly expected by society in order to solve global challenges and consequently establish a greater good for all (Benett et al. 2009). However, if businesses around the world performed their actions with a greater sense of consciousness, human flourishing and value creation for all could be achieved while solving the majority of global challenges in the twenty-first century (Sisodia et al. 2014). The following paragraphs will explain distinct characteristics, benefits, organizational practices, and perspectives which distinguish conscious business from traditional approaches.

2.3.1 Definition

Fred Kofman (2006), the author of *Conscious Business: How to Build Value through Values*, generally defines a conscious business as an organization which promotes mindfulness for all of its stakeholders. Moreover, it is based on humanity and fosters peace and happiness in individuals, respect and solidarity in the community, and mission accomplishment in the organization. Based on this understanding, the next section will outline key characteristics of conscious businesses.

2.3.2 Key Characteristics

The main task of a conscious business is to help people accomplish their mission while they develop healthy relationships and experience unconditional sense of peace, happiness, and growth by actualizing and transcending the self (Kofman 2006). Moreover, a conscious business continually examines itself in becoming a self-aware and responsible partner for its stakeholders, and thus purposely has a very low tolerance for dysfunctional tendencies (Renesch 2012). Ken Wilber (2006) states that conscious businesses focus on the dimensions mind, spirit, body, self, culture, and nature and unify them in an integral concept. Establishing such a business requires enacting the universal virtues of wisdom, courage, love, temperance, and transcendence (Kofman 2006). With regard to this, the term love refers to a selfless promotion of the growth of others who are impacted (Sanders 2008). Applying this mindset is likely to result in a maximization of social, spiritual, emotional, intellectual, ecological, and financial value for society as a whole. Moreover, people who interact with such a company experience more safety, trust, and fulfillment due to its systemic worldview (Sisodia et al. 2014).

2.3.3 *Organizational Characteristics*

Conscious businesses can be described as self-organizing systems which focus on ongoing organizational learning and self-actualization. A high degree of decentralization and collaboration enables such systems to adapt, innovate, and evolve faster in competitive environments. Focusing on unmet and evolving needs rather than existing demand is a key direction for conscious businesses to sustain competitive advantage (Mackey and Sisodia 2014). Focusing on functional and emotional stakeholder needs, offers conscious businesses a wider range of business opportunities than simply focusing on customer demands (Sisodia et al. 2014). Figure 2.1 summarizes key aspects of conscious business regarding its organizational development and orientation.

2.3.4 *Quantitative Indicators*

The practices and beliefs of the conscious business approach cannot be quantified entirely due to the level of complexity and the aspect that the linked attitudes do not reflect standardized measures (Mackey and Sisodia 2014). Nevertheless, there are several indicators which can be assessed to examine the outcomes of conscious business practices (Fig. 2.2). Therefore, this paragraph outlines a selection of measurable aspects.

Conscious businesses demonstrate a superior performance in terms of combining a stable and profitable financial development while realizing the needs of internal and external stakeholders. In terms of creating a sustainable financial development, conscious businesses usually create high sales numbers in combination with higher net margins compared to their traditional competitors. This development is a result of a strong acceptance by customers, committed employees, collaborative supplier relations, as well as a high level of organizational effectiveness (Mackey and Sisodia 2014). The sales performance of the wholesale company Costco has been steadily increasing over the last years. Costco generated $162 million per store in 2015, while maintaining moderate administrative expenses at 10% of their net sales. At the same time, the company has maintained its above-average wage

Fig. 2.1 Organizational development and orientation of conscious businesses (author's diagram)

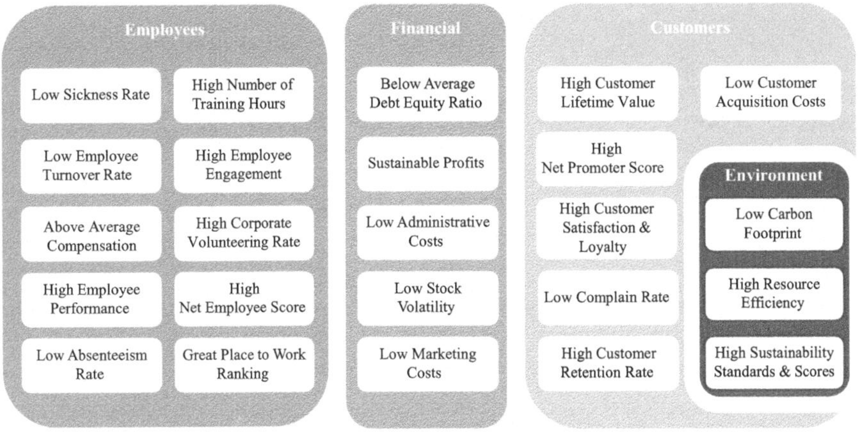

Fig. 2.2 Quantitative indicators of conscious businesses (author's diagram)

level for employees as well as collaborative pricing models with their suppliers (Costco 2015). Combining stakeholder interests with high profitability and productivity has been an essential aspect of Costco's success (Berman 2011).

Conscious businesses have been demonstrating that it is valuable and profitable to focus on a purpose beyond profit maximization, while including all important stakeholders. Investing in humanistic companies with a long-term perspective has proven to be valuable and rewarding (Sisodia et al. 2014). Sisodia et al. (2014) compared so called humanistic *firms of endearment* with the average performance of companies in the S&P 500 index as well as *Good to Great* companies which were described by Jim Collins (2001). The results of the comparison are shown in Fig. 2.3 with reference to timeframe 1998–2013.

The holding period returns were calculated by using closing prices and dividends of the respective stocks. The total return was calculated by accumulating the holding period returns to account for reinvested dividends. As a distinct factor, firms of endearment could achieve higher Price-Earnings Ratio (P/E) compared to the S&P 500 and Good to Great companies. The P/E ratio can be considered as an indicator for an integrated stakeholder approach as well as the level of sustained growth (Sisodia et al. 2014). However, the greatness of a company cannot be

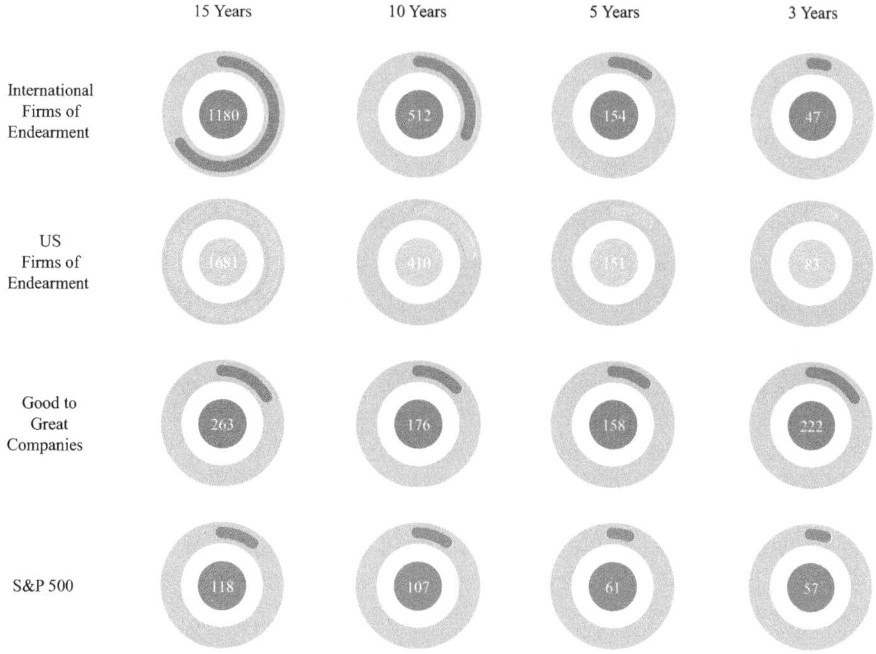

Fig. 2.3 Cumulative percental returns (author's diagram)

indicated by its financial performance alone but its quality of actions (Mourkogiannis et al. 2008). Maintaining superior performance is mostly based on the employee's performance and passion for their work. Hence, the development of individual abilities plays a significant role in constantly improving organizational effectiveness. Based on this perspective, conscious businesses invest in their employees by offering above-average training hours (Sisodia et al. 2014). Moreover, they invest in platforms and learning formats to give employees the possibility to align their passion and purpose with the company goals. Applying such person-centric mindset and individual empowerment is likely to result in better customer relationships due to an increased level of perceived support and personal appreciation. In addition, the underlying humanization of business builds the basis for a nurturing work environment that becomes manifest in low employee turnover rates, high employee commitment, collaborative stakeholder partnerships, and integrity throughout the organization (Sisodia et al. 2014). The retail chain The Container Store offers its full-time employees an average of 263 h of training per year compared to an industry average of 8 h. Moreover, the company fosters an employee centric work culture which is based on the development of individual abilities for their professional and private life. This attitude has been a significant success factor for the company in terms of their financial performance as well as their employee satisfaction. The turnover rate is less than 10% compared to an industry where employees usually change the workplace in less than 1 year (The Container Store 2016).

In terms of productivity, The Container Store views one great employee equal to three good employees and hence pays an above-average minimum salary of $50,000 to each full-time employee (Tindell 2011). 98% of their personnel state that the company established a great work atmosphere and 95% outlined that the Container Store cares strongly about their personal needs. Overall, 91% of their workforce regards the Container Store as a great place to work (Great Place to Work 2016). With reference to this, Mackey and Sisodia (2014) add that the great place to work ranking is a good proxy for human-centric and conscious businesses in general. Another distinct indicator for conscious businesses can be found in their diminished reliance on classical marketing approaches. Their marketing approach differs as they create and maintain their customer base by direct customer experience and word-of-mouth rather than spending a high budget on advertising. This approach results in lower overall marketing costs, while keeping a clear strategy on how to attract and retain customers (Sisodia et al. 2014). Thus, the strategy emphasizes the design of impactful, sustainable and mission-driven marketing instead of focusing on quantitative-driven marketing. Applying *sense-of-mission marketing* strengthens purpose-driven brands and also results in a higher internal cohesion, a clear company direction and signals the company's long-term sustainable orientation. Patagonia is pursuing such a marketing strategy by focusing their marketing approaches to the company's mission (Armstrong et al. 2015). Patagonia has the mission to build the best product, cause no unnecessary harm, use business to inspire, and implement solutions for the environmental crisis (Patagonia 2016a).

In 2011, Patagonia launched a campaign called *Don't buy this jacket* as a contrast to the busiest online shopping day *Cyber Monday*, which was introduced by the National Retail Federation. In this campaign, Patagonia raises awareness regarding the consumption of natural resources in production processes, asks consumers to buy less and consciously reflect on buying decisions in order to preserve an inhabitable world for future generations. The campaign addressed the Patagonia *common threads initiative* which outlines the positive impact through reduction, repair, reuse and recycling of products. The combination of the mentioned aspects above builds the basis to reimagine a world where humanity only takes what nature is able to replace (Patagonia 2011). In 2013, Patagonia launched its *Worn Wear* campaign to further raise awareness on resource consumption and possibilities to prevent unnecessary CO_2 emissions and waste. The company employs 45 full-time employees in Reno, Nevada who complete around 40,000 repairs per year (Patagonia 2016c). Moreover, Patagonia collaborates with *iFixit* to create care and repair guides to educate and empower people (Patagonia 2016b). In order to involve customers into this process, Patagonia started the international *Mobile Worn Wear Tour* by driving with a biodiesel repair truck to different locations and offering on-site repair services for used articles (Simpson 2015).

2.3.5 Conscious Employees

As the previous paragraphs outlined the organizational perspective regarding conscious business practices, this chapter focuses on the attitudes which characterize conscious employees. Fred Kofman (2010), views conscious employees as an organization's most important asset, whereas unconscious employees embody its most dangerous liability. Conscious employees align their individual purpose with the company's purpose and take responsibility for acting accordingly (Tate 2015). Investigating the world with rigorous scientific reasoning, and applying a high level of self-reflection with equally moral reasoning, is another characteristic according to Kofman (2006). Conscious employees do not compromise human values for material success and act with honesty, integrity, and respect regarding their environment. Moreover, they contemplate their own self-conception in terms of creating a virtuous, meaningful, and happy life for themselves while accepting other perspectives and colleagues in a non-judgmental manner. Nevertheless, they are able to productively express their emotions and seek mutually beneficial solutions in disagreements (Kofman 2006). Figure 2.4 summarizes the key characteristics of conscious versus unconscious businesses attitudes, behaviors, and responses (Kofman 2013a):

Emotional mastery is required as a prerequisite for the mentioned parts above, as all of them involve an emotional dimension and interpersonal interactions. Being able to control emotions in a business environment has important implications towards the consciousness of decisions, judgments, assumptions, predictions, interpretations and more. Managing conflict situations with stakeholders, leading the

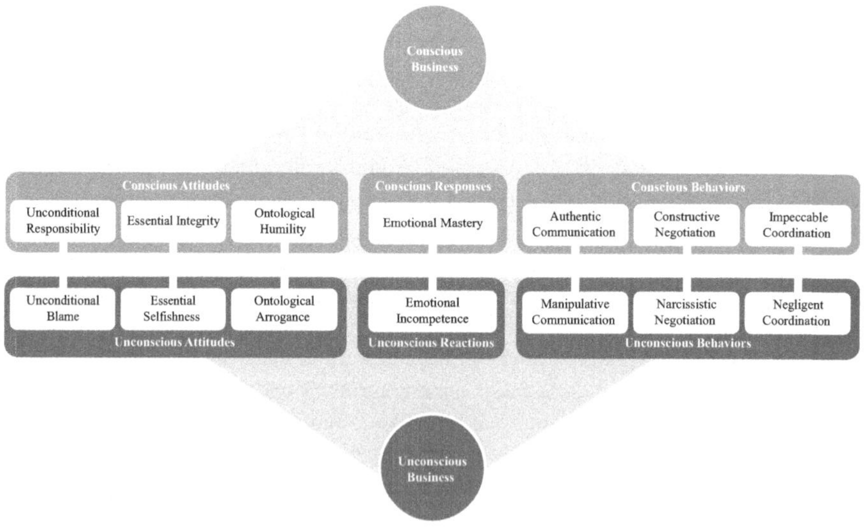

Fig. 2.4 Conscious business versus unconscious business (author's diagram)

company through critical states, or fostering a transformation process commonly share this high level of emotional involvement (Kofman 2006).

2.4 Related Concepts

This chapter gives an overview regarding related concepts to conscious business. The following concepts were chosen to highlight differences and similarities to existent and established practices.

2.4.1 Corporate Social Responsibility

According to the World Business Council for Sustainable Development (1998), corporate social responsibility refers to a continuing commitment by business to behave ethically and contribute to economic development, while improving the quality of life of the workforce, their families, local community, and society at large. The responsibility arises from this definition is combined in a triple bottom line approach, which combines an environmental, societal, and economical perspective (WBCSD 1998) The European Commission (2011) adds that this commitment goes beyond legal obligations with regard to society and the environment on a voluntary basis. However, the triple bottom line is missing passion and purpose to reflect humanity in business and thus enable companies to conduct their business in a sustainable way (Poon 2010). Idowu et al. (2014) outline two essential approaches that companies apply with reference to CSR. The concept is either carried out in a *before-profit* or *after-profit* manner. Companies which use the first approach implement the framework in everything they do, while others implement CSR as an add-on to their business practices and thus do not practice it as an integral part of their business.

Mackey and Sisodia (2014) view CSR as a fallacy as it indicates a tainted or ethically neutral structure of business. In addition, Porter and Kramer (2006) view CSR efforts often as counterproductive regarding its outcomes. First, they pit business against society, when in reality the two are interdependent. Second, they pressure companies to think of corporate social responsibility in generic ways instead of in the way most appropriate to their individual strategies (Porter and Kramer 2002). Mackey and Sisodia (2014) draw a clear line between CSR and conscious business. From a CSR perspective, essential elements such as a company's purpose and culture are not reflected which embodies an integral part of conscious businesses. Whereas conscious businesses see themselves as a part in a complex and adaptive system, the CSR point of view offers a mechanistic view paired with a trade-off mindset regarding stakeholder interest. Although CSR can be an inherent part of a company, Mackey and Sisodia (2014) remark that the related measures are usually performed by a separate department with a focus on communicating charitable gestures rather than critically reflecting on the company's impact. This understanding is often driven by a traditional leadership understanding which encompasses a limited overlap between business and society. In contrast, conscious leaders see responsible behavior as a core of the company's foundation and recognize that business is a mutually dependent relationship to society.

2.4.2 B Corporations

B Corporations aim at harnessing the power of purpose-driven enterprises to create public benefit. Moreover, the approach is focused on aligning various stakeholder interests with business interests. In 2006, a non-profit organization was founded under the name *B Lab*. The initiators created a set of performance and legal requirements with regard to the aspects mentioned above (Honeyman 2014). The founders of B Lab envision in their Declaration of Interdependence, a global economy that uses business as a force for good (B Lab 2016a). B Corporations embody socially-aware enterprises and thus integrate social benefit as a core mission (Lawrence and Weber 2011). Non-profit organizations as well as government agencies are not eligible for becoming a B Corporation. In 2007, the first company was officially certified as a B Corporation. The certification process focuses on a defined standard of social and environmental performance, accountability, and transparency in companies. The B Corporations movement wants to initiate a change from shareholder maximization to shared value maximization. Companies that consider a B Corporation certification need to assess their social and environmental performance by conducting the B Lab *Impact Assessment*. This evaluation is a free online-based tool in the areas of governance, workers, community, environment, and the respective impact, which is represented the company's business model (Honeyman 2014).

The B Corp Impact is split into five dimensions with multiple items (Table 2.1).

All factors are first assessed by answering a set of questions which are evaluated at the end on a 200-point scale (Honeyman 2014). As a next process step, applying companies need to provide supporting documentation and schedule an assessment review with members of the B Lab. If the respective company receives a score above 80, it receives the B Corporation certification for 2 years before they can apply for a re-certification (B Lab 2016b). A certification fee applies to the process according to the company revenue and ranges from 500 to 50,000 € (Gexsi 2016). Currently, more than 1780 businesses in 50 countries and 130 different industries are certified as B Corporations. Patagonia, Ben & Jerry's, Etsy and many more joined the B Corporation community (B Lab 2016c). In 2015, Paul Polman, the CEO of Unilever announces Unilever's participation and support in the B Lab's Advisory Council for Multinational Companies to scale the movement especially for multinational companies (Polman 2015). Mackey and Sisodia (2014) point out that B Corporations are compatible with the four tenets of conscious businesses,

Table 2.1 B Corp impact assessment (author's table)

Dimension/item	Company score	Median score
Environment		
Environmental Products & Services		
Environmental Practices		
Land, Office, Plant		
Energy, Water, Materials		
Emissions, Water, Waste		
Suppliers & Transportation		
Workers		
Compensation, Benefits & Training		
Worker Ownership		
Work Environment		
Customers		
Customer Products & Services		
Products & Services		
Serving Those in Need		
Community		
Community Practices		
Suppliers & Distributors		
Local		
Diversity		
Job Creation		
Civic Engagement & Giving		
Governance		
Accountability		
Transparency		
Overall		

higher purpose, *conscious leadership*, *conscious culture* and *stakeholder integration*. However, they criticize that this model, disempowers owners regarding their control over the business, as management decides about measures in the B Corporations governance model.

2.4.3 Economy for the Common Good

In 2001, Joachim Sikora and Günter Hoffman initiated the first vision for the *Economy for the Common Good* or *Gemeinwohl-Ökonomie*.

The vision, to realign the existing economic system towards a more human, ecological, and democratic form was then made concrete by the Austrian, Christian Felber in 2006. In cooperation with members of different institutions and companies, the Economy for the Common good movement was established in Vienna four years later (Gemeinwohl-Ökonomie 2016a). The Economy for the Common Good can be classified as an ethical market economy (Felber 2015). The core of the Economy for the Common Good lies in its primary orientation of economic actions towards the maximization of common good, rather than wealth maximization for a limited group of individuals. The movement aims at integrating business and society on the basis of collaboration, solidarity, mutual appreciation, and shared values which do not act in a conflicting or exclusive manner (Felber 2014). Moreover, the Economy for the Common Good implicates a redefinition of economic success indicators and metrics such as the GDP on a macro-economic level and profit metrics on business level. This redefinition tries to combine monetary with non-monetary aspects such as the perceived quality of life and others (Felber 2012). An additional aim can be identified in changing the notion from harmful competition to collaboration on macro and micro-economic levels (Felber 2014). Felber (2014) acknowledges the creation of profits as a necessary aspect of business but considers it as means to an end to establish a common good. In order to measure the contribution of a company to the common good, a matrix has been established to map the dimensions *Human Dignity*, *Cooperation and Solidarity*, *Ecological Sustainability*, *Social Justice*, and *Democratic Co-determination and Transparency*. The matrix measures how these aspects are lived in support of the impacted stakeholder groups of a company. Establishing this matrix as an effective reporting tool is based on the premise of accountability, holism, measurability, comparability, comprehensibility, public accessibility, external auditing, and its legal implications. With regard to the certification process, a team of external common good auditors review and rate each dimension. If a company does not fulfill certain criteria or acts in a non-beneficial way, negative points will be added at the bottom of the matrix (Felber 2014). An exemplary matrix is outlined in Fig. 2.5.

The auditors consider 52 individual items on the five dimensions and 17 items for potential negative points (Gemeinwohl-Ökonomie 2016c). Based on the overall score, companies should be granted an adjusted tax rate, tariffs, credit rates, and priorities in public bidding processes, research agreements, direct investments and other criteria. Hence, the company score should be correlated with granted benefits by political and societal institutions (Felber 2014). At the moment, the Economy for the Common Good certification embodies a voluntary initiative. However, the European Economic and Social Committee (EESC) stated that it aims to include the principles on a European and national dimension. Moreover, the Economy for the Common Good is considered as beneficial to the Europe-2020-Strategy to establish an ethical market economy (EESC 2015). Currently, more than 2000 companies have been certified by the described evaluation process. Most certified companies operate in Germany, Austria and Switzerland but there is also a growing number is businesses in other European countries such as Italy, the Netherlands, Sweden, Serbia, and others (Gemeinwohl-Ökonomie 2016b). In addition, regional groups and think tanks in North and South America are supporting the development of this movement (Felber 2014).

COMMON GOOD MATRIX 4.1

This version is valid for Common Good Balance Sheets generated in 2013

ECONOMY FOR THE COMMON GOOD

An economic model for the future

VALUE / STAKEHOLDER	Human dignity	Cooperation and Solidarity	Ecological Sustainability	Social Justice	Democratic Co-determination and Transparency
A) Suppliers	**A1: Ethical Supply Management** Active examination of the risks of purchased goods and services, consideration of the social and ecological aspects of suppliers and service partners				90
B) Investors	**B1: Ethical Financial Management** Consideration of social and ecological aspects when choosing financial services; common good-oriented investments and financing				30
C) Employees, including business owners	**C1: Workplace quality and affirmative action** Employee-oriented organizational culture and structure, fair employment and payment policies, workplace health and safety, work-life balance, flexible work hours, equal opportunity and diversity 90	**C2: Just distribution of labor** Reduction of overtime, eliminating unpaid overtime, reduction of total work hours, contribution to the reduction of unemployment 50	**C3: Promotion of environmentally friendly behavior of employees** ctive promotion of sustainable life-styles of employees (mobility, nutrition), training and awareness-raising activities, sustainable organizational culture 30	**C4: Just income distribution** Low income disparity within a company, compliance with minimum and maximum wages 60	**C5: Corporate democracy and transparency** Comprehensive transparency within the company, election of managers by employees, democratic decision-making on fundamental strategic issues, transfer of property to employees 90
D) Customers / Products / Services / Business Partners	**D1: Ethical customer relations** Ethical business relations with customers, customer orientation and co-determination, joint product development, high quality of service, high product transparency 50	**D2: Cooperation with businesses in same field** Transfer of know-how, personnel, contracts and interest-free loans to other business in the same field, participation in cooperative marketing activities and crisis management 70	**D3: Ecological design of products and services** Offering of ecologically superior products/services; awareness raising programmes, consideration of ecological aspects when choosing customer target groups 90	**D4: Socially oriented design of products and services** Information, products and services for disadvantaged groups, support for value-oriented market structures 30	**D5: Raising social and ecological standards** Exemplary business behavior, development of higher standards with businesses in the same field, lobbying 30
E) Social Environment: Region, electorate, future generations, civil society, fellow human beings, animals and plants	**E1: Value and social impact of products and services** Products and services fulfill basic human needs or serve humankind, society or the environment 90	**E2: Contribution to the local community** Mutual support and cooperation through financial resources, services, products, logistics, time, know-how, knowledge, contacts, influence 40	**E3: Reduction of environmental impact** Reduction of environmental effects towards a sustainable level, resources, energy, climate, emissions, waste etc. 70	**E4: Investing profits for the Common Good** Reducing or eliminating dividend payments to extern, payouts to employees, increasing equity, social-ecological investments 60	**E5: Social transparency and co-determination,** Common good and sustainability reports, participation in decision-making by local stakeholders and NGO's 30
Negative Criteria	Violation of ILO norms (international labor standards) / human rights -200 Products detrimental to human dignity and human rights (e.g. landmines, nuclear power, GMO's) -200 Outsourcing to or cooperation with companies which violate human dignity -150	Hostile takeover -200 Blocking patents -100 Dumping Prices -200	Massive environmental pollution -200 Gross violation of environmental standards -200 Planned obsolescence (short lifespan of products) -100	Unequal pay for women and men -200 Job cuts or moving jobs overseas despite having made a profit -150 Subsidiaries in tax havens -200 Equity yield rate > 10 % -200	Non-disclosure of subsidiaries -100 Prohibition of a works council -150 Non-disclosure of payments to lobbyists -200 Excessive income inequality within a business -150

Fig. 2.5 Economy for the common good matrix (Economy for the Common Good 2016)

Table 2.2 Comparison of concepts (author's table)

Dimension		Initiative: CSR	B Corporations	Economy for the Common Good	Conscious Business
Humanistic Orientation			•	•	•
Internal	Purpose		•	•	•
	Leadership				•
	Culture			•	•
	Employees	•	•	•	•
	Profit	•	•	•	•
External	Society	•	•	•	•
	Environment	•	•	•	•
	Investors		•	•	•
	Customers		•	•	•
	Suppliers		•	•	•
	Competitors			•	•
Number of Covered Dimensions		4	9	11	12
Reporting Standards		•	•	•	
Certification			•	•	

2.4.4 Comparison of Concepts

Table 2.2 displays a comparison of conscious business, the concepts of a triple-bottom-line CSR, B Corporations, and the Economy for the Common Good. The dimensions are split into the coverage of internal and external stakeholders and motives as well as their humanistic orientation.

2.5 Economical Drivers for Change

This chapter examines economical drivers, key elements and further aspects which are related to the emergence of conscious business.

The following paragraph will highlight main characteristics of capitalistic economic systems as this system represents the environment in which conscious businesses operate.

2.5.1 Traditional Capitalism

The *Wealth of Nations* by Adam Smith can be considered as the origin of Western economic thought (Baptist and Hyman 2014). Adam Smith postulated the existence of an *invisible hand,* which shapes market outcomes by *reasoned* decisions of people. Nevertheless, Adam Smith noted the importance of involved emotions in economic decisions in his work *The Theory of Moral Sentiments*. Unfortunately, the foundations of capitalism in the past two centuries have been predominantly based on only the first pillar (Sisodia et al. 2014). Kromphardt (2015) outlines that there are three main dimension by which economic systems can be classified as *decision processes*, *coordination*, and *underlying motivation*. However, these aspects can only be considered as a basic framework and do not reflect the complexity of each system. In capitalistic systems, the planning and decision process concerning production, distribution and consumptions is predominantly driven by individual economic systems. These economic systems are characterized by private ownership of capital. The separation between a relatively small group with a high degree of decision-making power, and a larger group without a considerable level of ownership regarding production factors is a distinct characteristic in capitalistic systems. Both groups participate in a market where the second group is able to decide for a transaction partner in terms of employment and consumption. Supply and demand are generally coordinated as a dynamic process between both groups and thus decisions are coordinated in terms of price, quantity, and further related criteria (Kromphardt 2015). The underlying motivation in capitalistic systems can be described as predominantly rational and individual with a focus on profitability for capital-owners (Zimbalist and Sherman 1984). The emergence of *financial capitalism* in the twentieth century and a connected, unequal distribution of wealth were strongly driven by the idea of shareholder value maximization. Pursuing short-term profitability in favor of a company's shareholders has forced most businesses to view this aim with the highest priority (Kromphardt 2015). Concerning the social responsibility of business in capitalistic systems, Milton Friedman (1970) postulated that the increase of profits embodies the only social responsibility while engaging in free competition without deception or fraud. Mourkogiannis et al. (2008) disagree with this perspective, as pure profit maximization is neither valid nor ethical as a sole aim in social responsibility. They argue that a company is not only existent because of its shareholders, but because of a complex web of stakeholders. Hence, the profit maximization perspective neglects the nature of business. Drucker (2008a) adds to this characteristic, that profits cannot be taken as a cause or rationale of business behavior and business decisions, but rather the test of their validity. Nevertheless, the idea of business's sole purpose of profit maximization is unlikely to disappear soon (Sisodia et al. 2014). With reference to this outlook, Gates (2008) highlights the ability of capitalism to make self-interest serve a collective societal interest.

A systemic change is needed, where market incentives including profits, are designed to ease the world inequities (Gates 2008). The underlining market ideology therefore has to switch from a *demand* to a *need* perspective to contribute to societal goals (Kinsley 2008).

2.5.2 Capitalistic Systems in the United States and Germany

Conscious business first evolved as a concept in the United States. Hence, this paragraph examines major differences between the surrounding economic systems in the United States of America and in Germany with regard to business environments. Classifying different national capitalistic systems can be done by comparing the manifestation of liberal versus coordinated regulatory approaches. Nevertheless, national capitalistic systems usually combine elements of both approaches in different areas, such as corporate management, taxation, corporate finance, social security systems, education, labor relations and other. Hence, most countries cannot be identified as purely liberal or coordinated. The capitalistic system in the United States of America in classified as predominantly liberal. In comparison to coordinated approaches, the American system favors regulations by markets rather than by central governmental organizations. As a result of an individualistic orientation, collectively regulated labor agreements are usually only established on a corporate level and not for an entire profession (Schröder 2014). A survey conducted by GlobeScan in 2009 revealed that 53% of Americans agree that capitalism creates problems which can be solved through reforms and regulations. About 25% state that capitalism in the United States works well and efforts to reform will result in inefficiencies and 13% say that capitalism needs to be replaced by a new economic system (GlobeScan 2009). In contrast, the German economic system is a combination of a market-driven capitalism with a strong focus of income redistribution for supporting a large social safety net (Schröder 2014). Economies which link these aspects are commonly described as *social market economies* (Rosser and Rosser 2004). Based on a GlobeScan study (2009, 8), 75% of Germans believe that capitalism has issues that can be solved by applying regulations and reforms. About 16% state that capitalism works well in their country and 8% pledge for a different economic system. A majority of the German population believes that business profits are a necessary prerequisite in order to establish societal welfare. Nevertheless, only 22% perceive the distribution of these profits to be just (Baumann and Schulz 2015). Brandt and Georgiou (2016) conclude that American businesses tend to incorporate *shareholder-capitalism* whereas German businesses can be considered to follow the direction of *stakeholder-capitalism*. Based on the idea of the social market economy, German businesses consider the integration of various stakeholder groups and societal welfare with a higher priority than most American businesses do (Wellner 2001). This mindset is also visible in different motives of investors and business owners. According to Abelshauser (2011), investors in Germany predominantly follow the motive of sustainability and a long-term value maximization, rather than pursuing short-term profit gains. Especially small and medium sized businesses in Germany embody this philosophy. In 2013, 99.3% of all businesses in Germany were categorized as small and medium sizes businesses which therefore mark the most frequent type of organizations (Statistisches Bundesamt 2013). In general, a symbiosis of state, business, and society with a focus on common goals has been an enduring characteristic of the social market economy in Germany in comparison to other neo-liberal forms of capitalism.

The financial crisis in the early 2000s can be considered a systemic crisis of a deregulated financial capitalism which found its origin in the United States and still sparks the discussion, which form of market-based capitalism benefits society at large (Abelshauser 2011).

2.5.3 Conscious Capitalism

The nature of capitalism has to be transformed in order to regain trust among societies worldwide. People around the globe are not willing to accept the logic of shareholder maximization at the expense of society and the environment anymore. Economic and societal prosperity and well-being need a liberal and democratic market-based framework. As other economic systems, such as socialism and communism, do not represent a valuable alternative, capitalism has to be transformed in ways which benefit all stakeholders (Schwab 2016). Sustainable value creation is a joint commitment of stakeholders to satisfy their needs and desires. Hence, stakeholder capitalism allows a humanization of economic behavior and represents the most suitable way to allow societal and economic advancement for business and society (Freeman 2002). In 2011, the World Economic Forum (2011) conducted a study among 14,300 people regarding the role of values in the post-crisis economy. More than 67% of the surveyed people answered that this crisis has been also a crisis of values. In Germany, more than 63% share this perspective. Asked about the most important values in the global political and economic system, 39% view *honesty, integrity, and transparency* (Germany: 38%) as the most important elements, followed by *respecting others' rights, dignity, and views* (Global: 24%/Germany: 26%), *preserving the environment* (Global: 17%/ Germany: 18%), and the *impact of actions on the well-being of others* (Global: 20%/Germany: 18%). Aburdene (2005) outlines a needed transformation of capitalism as a result of two different processes. Accordingly, top-down events such as rising deficits, accounting scandals, rising inequalities, misconduct of corporate executives, and a lack of ethical behavior account for the first dimension. In addition, changing societal values such as a demand for just wealth allocation, socially responsible investing, and more participation are examples of bottom-up events. Mackey and Sisodia (2014) add that the focus on self-interest and profit maxims have to be overcome as well to create the ground for an ethically compelling version of capitalism. Overall, these changing expectations are also a result of a search for meaning in the current economic system and will consequently change the very soul of capitalism. Sisodia et al. (2014) call it a *social transformation* of capitalism as a transcendent shift in the *zeitgeist* of contemporary society. This is mainly influenced by a growing number of people in predominantly Western societies, who did not experience positive benefits of the prevalent capitalistic system. However, this social transformation is linked to a high level of uncertainty for both individuals and businesses regarding the future economical and societal development (Aburdene 2005). In conclusion, Aburdene (2005) views this development as the validity for a rising consciousness in capitalism. Conscious Capitalism aims at restoring damaged trust and acceptance by recapturing the purpose of business as a collective value creator and thus improves the well-being of society on an economic, social, intellectual, ecological, and spiritual dimension (Mackey and Sisodia 2014). Mackey (2016) describes Conscious Capitalism as a more complex form of capitalism that reflects and leverages the interdependent nature of life and all of the stakeholders in a business.

It embodies an evolving paradigm which is focused on creating a higher collective good and thus a more harmonious connection between business and society (Mackey and Sisodia 2014). "Together, business leaders can liberate the extraordinary power of business and capitalism to create a world in which all people live lives full of purpose, love, and creativity—a world of compassion, freedom, and prosperity. This is our vision for Conscious Capitalism" (Mackey and Sisodia 2014, p. 32).

Mackey and Sisodia (2014) emphasize that Conscious Capitalism is a market-based system which centers around the integration of a higher purpose to elevate humanity and build the base for every company's justification to exist. Conscious Capitalism does not reflect an entire framework on a macro economical level. However, it emphasizes the creative power of businesses as the driving source to positively transform the current economic system and thus rebuild societal trust and acceptance. With regard to the role of businesses, Kaku (1997) outlines that companies control vast resources around the globe employ millions of people and create an incredible amount of wealth. As an example, by the end of 2014, the 500 most valuable companies account for almost $32.3 trillion in terms of market capitalization, had $106.2 trillion in total assets, generated a net income of more than $2 trillion and employed around 46.1 million people worldwide (Financial Times 2015). If businesses unite their creative power in a more mindful and impactful way rather than pursuing wealth accumulation, the world would likely experience a higher level of collective prosperity, sustainability, and peace (Kaku 1997). Establishing socially conscious citizenship by corporations at scale would be beneficial for all stakeholders and strengthen the reputation for businesses as well as the surrounding economic system as a whole (Shaich 2011).

2.6 Societal Drivers for Change

This paragraph discusses major societal drivers for change towards increasingly conscious work environments as well as societies at large. The chapter encompasses indicators regarding value changes in society, the workplace, changes in information networks, and consumer behavior.

2.6.1 Value Changes in Society

"The most striking feature of contemporary culture is the unslaked craving for transcendence" (Delbanco 1999, p. 113). Sisodia et al. (2014) outline that most developed societies have entered the *Age of Transcendence* which signifies a turning point where physical (materialistic) influences strongly decrease in importance in favor of metaphysical (experimental) influences. In addition, a shift in the foundation from an objective base to a subjective base is increasingly prevalent both in private and professional environments. In the Age of Transcendence, subjectivity can be considered a primary factor in terms of establishing stakeholder loyalty and affection. More and more companies try to cope with this development by shifting their perspective from a rather rational to an emotional essence regarding their stakeholder relationships, especially with customers. On the other side, consumers increasingly tend to base their buying decisions on the emotional experience rather than focusing on the functional character of products and services. In terms of their work environment, a growing number of people evaluate their current and potential workplace regarding the fulfillment of social and emotional needs as well as its meaningfulness rather than judging it by the monetary income alone (Sisodia et al. 2014).

This development is closely linked to a significant change of values during the last decade. Self-actualization has increasingly grown in importance whereas the accumulation of wealth and ownership tend to be less important. This change also offers an explanation for a stronger orientation towards individualism and pluralization of social groups and lifestyles (Müller 2012). Overall, the growing influence of self-actualization on mainstream culture is reshaping business environments and societies.

Bakan (2004) adds that this development can be understood as a *self-actualization of capitalism* where also a growing number of companies shift their corporate egos and focus increasingly on the needs of their stakeholders including society. Sisodia et al. (2014) point out that these societal shifts are mainly driven by an aging society and their increasing prevalence of midlife values. With reference to the work of the psychologist and philosopher Erich Fromm, Sisodia et al. (2014) see a transition from a self-centered and materialistic *having society* to an others-centered *being society* which is deeply vested in high-ground moral values. Bennis (2014) notes that the growing population above the age of 40 and the connected psychological maturity is a major driver of this systemic change. Although, the preference for material possession will not disappear, a shift in the consumer behavior towards less ownership and to a higher accessibility and usage of goods and services can be clearly observed (Sisodia et al. 2014). This transition is also linked to the concept of *generativity* by the Swedish psychologist Erik Erikson. Generativity describes a virtuous behavior which is predominantly manifest in midlife. This behavior is characterized by the will to responsibly guide the next generation in a positive and productive way. Moreover, it is based the virtues of caring, giving, sharing, nurturing, and developing (Lerner et al. 2003). Kaplan

(2016) outlines that this concept is a significant aspect of individuals from 40 to 65 years and is guided by the will to pursue a worthwhile and meaningful life. In general, "generativity motivates behavior aimed at promoting the greater good" (Lachman 2001, p. 402). This concept is also closely related to the idea of *sustainability*. Sustainable development is defined as one that "meets the needs of the present without compromising the ability of future generations to meet their own needs" (World Commission on Environment and Development 1987, p. 41). Taking this line of argument into account, the distribution of the population in Germany currently shows a median age of 45.5 years (Fig. 2.6).

In 2030, the median age is predicted to rise to 47.1 (with a continuous migration trend) and 47.6 (with a lower than expected migration) (Fig. 2.7).

Sisodia et al. (2014, p. 220) conclude that "the collective maturation of especially western societies is driven by an aging population and rapidly advancing

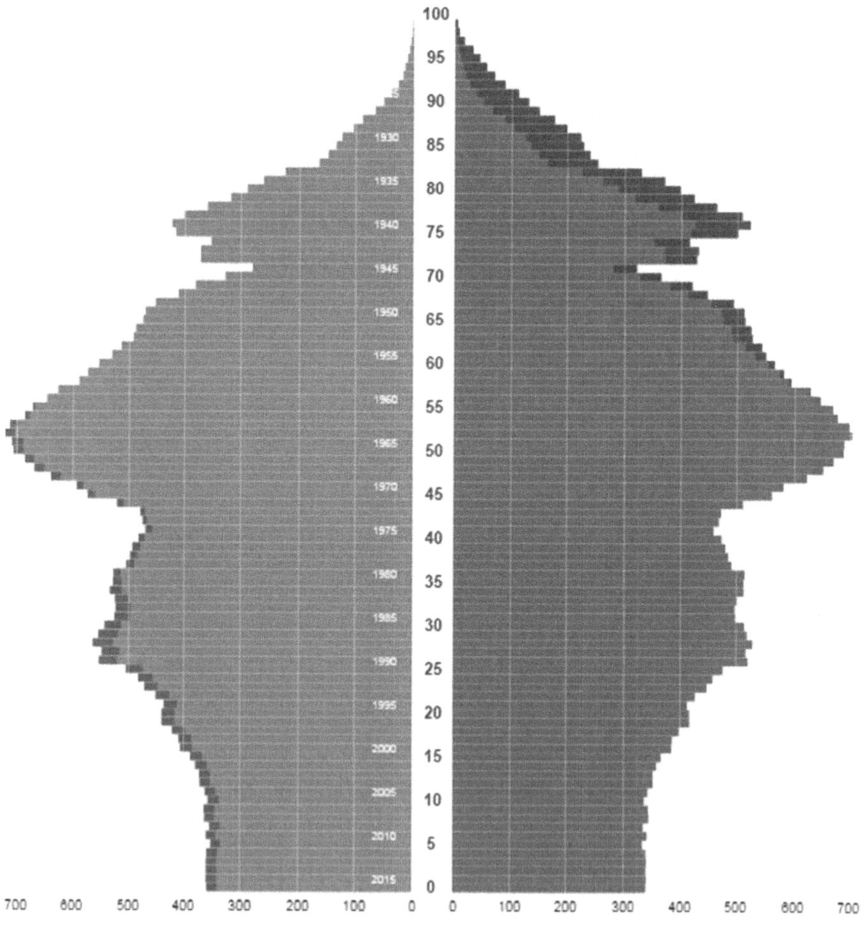

Fig. 2.6 Population pyramid in Germany 2016 (Statistisches Bundesamt 2016)

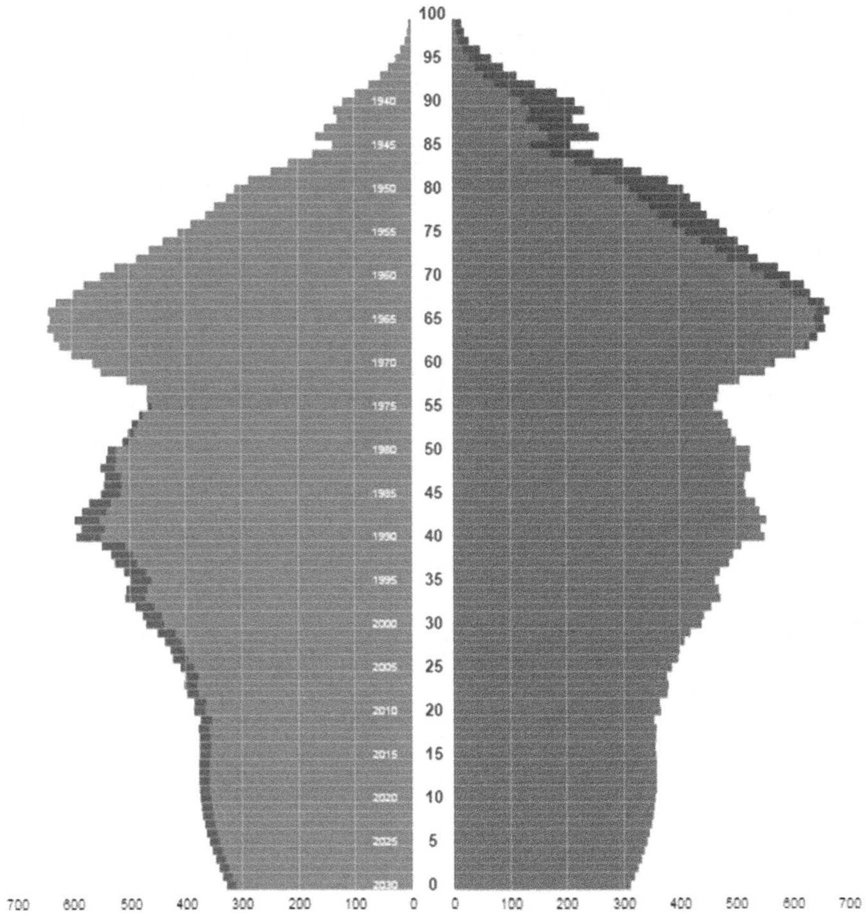

Fig. 2.7 Assumed population pyramid in Germany 2030 (Statistisches Bundesamt 2016)

human consciousness." Although value changes and beliefs differ among age groups, the trend towards a collective individualism represents a common element regarding society in Germany (Opaschowski 2010). Moreover, a focus on honesty and integrity can be identified as an essential element across generations. Among all values, *safety* represents the most dominant value in the German society. The GfK Association (2016) outlines, that this fact is a result of an increasing perception of political and financial instability in the past years. Fifty percent of the survey respondents see an increasing importance in the value *achievement*, 49% *solidarity*, 48% *home*, 47% *trust*, 44% *responsibility*, and 43% in *optimism*. Only 38% named the value *competition* and 34% the value *power* as increasingly important for the future. With regard to the age group from 14 to 24, *innovation* was named as the second most important value (GfK 2016). Corfe (2008) indicates an ever-extending human rights movement as an additional indicator concerning the emergence of more conscious societies in developed countries. Hence, fostering economic and emotional welfare acts as a shared common purpose in a responsible and meaningful society. With reference to the spiritual values, Corfe (2008, p. 64) states that spirituality is not merely based on an institutional religious understanding but "the belief in the infinite good coming into being." Nefiodow and Nefiodow (2014) view the creation of spiritual health as a major aspect in achieving holistic health for human beings. Holistic health is characterized as the driving force behind the 6th Kondratieff cycle which encompasses the current and future economic development. Figure 2.8 emphasizes the past and predicted Kondratieff waves as well as their driving factors.

The emergence of psychosocial health includes the physical, psychological, ecological, spiritual, and social dimension of human health and embodies the basic driver of a new long wave in the economy. Psychosocial health refers to vital relationships and well-being of individuals regarding themselves as well as others. Nefiodow and Nefiodow (2014) view the development of individual and collective consciousness as the link to achieving psychosocial health. "The sixth Kondratieff will be the first long wave, whose focal point won't be investments in hardware, machines or chemical process, but rather the human being with his/her mental and social problems, needs and potential" (Nefiodow and Nefiodow 2014, p. 62).

Psychosocial health has direct implications towards the quality and quantity of economic development as it is linked to the overall effectiveness of work and the productivity level. Allianz Global Investors (2010) perceive holistic health and the attached economic opportunities as a megatrend and a growing market in terms of products and services. The OECD (2013, p. 6) expects that on "average across OECD countries, total health and long-term care expenditure is projected to increase by 3.3 and 7.7 percentage points of GDP between 2010 and 2060". In 2014, the overall health expenditure accounted for 9.9% of the world GDP. In Germany, this expenditure rose from 9.4 in 1995 to 11.3 in 2014. In the same time

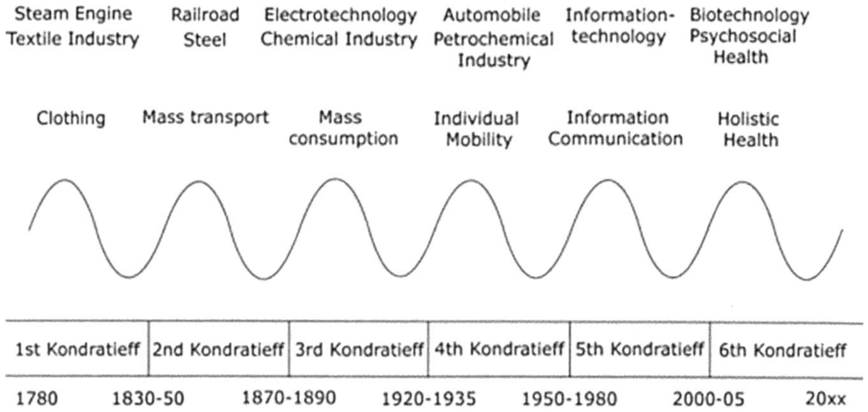

Fig. 2.8 Kondratieff waves (Nefiodow and Nefiodow 2014, p. 59)

period the United States experienced an increase from 13.1 to 17.1% of the national GDP (World Bank 2016a). Establishing psychosocial health represents a global need and therefore calls for global strategies in an increasingly interconnected world. Coping with the emergence of psychosocial health also requires businesses to establish conditions under which holistic health can be fostered as the well-being of employees has direct impact on their overall performance and effectiveness (Nefiodow and Nefiodow 2014). A study conducted by the Forsa Institute (2014) points out that only 25% of the participants perceived their work as a positive contribution whereas 42% pronounced a negative influence on their health and well-being. With regard to this, Mackey and Sisodia (2014) note that a large number of businesses still do not consider the well-being of its members with a high priority and thus do not reflect this change of a rising consciousness in society.

2.6.2 Workplace Values

The job embodies an essential part in the life of Germans. Whereas *family and relationships* were named as the most essential part of life by 39%, the *job* occupies the primary element for 34%, followed by leisure with 20%, and civic engagement with 7% (Bertelsmann Stiftung 2015). As outlined in Sect. 2.5.1, safety represents the most important value for Germans. This result can also be identified with reference to the workplace. An Orizon (2014) survey among more than 2000 participants also came to the conclusion that *job security* represents the most important factor among all age groups. *Fair compensation*, *flexible work hours*, *workplace proximity*, and *task variation* represent the ranks 2–5. Although monetary income represents a source of motivation for employees in Germany, it does not represent the guiding principle for work itself if basic needs are fulfilled. This perspective is especially manifest in the age group from 18 to 29 (Bertelsmann Stiftung 2015). This study outlined that, in addition to job security, the perceived *independence and self-determination* as well as the *contribution of individual ideas* play an important role with regard to workplace expectations. *Above-average income* has been named only as the fifth most important criteria. Furthermore, establishing *balance of private and professional* life is valued with an increasing importance among future expectations. Mackey and Sisodia (2014) identified the emergence of the Millennial generation in the work environment as an additional reason towards a more conscious shift in business and society. Millennials are also referred to as the Generation Y and have been named first under this classification by Strauss and Howe (1992) and are referred to as individuals who are born in the mid-1980s. Whereas the research institute Gallup (2016) defines this cohort as being born from 1980 to 1996, the Pew Research Center (2016) applies the time period from 1981 to 1997. As there is no unified understanding of a distinct time period, this book will apply the classification by the Pew Research Center.

Mackey and Sisodia (2014) outline that a classifying element of the Millennial's motivation is their pursuit of a higher purpose at work which is closely linked to the aim of living a meaningful life. Benett et al. (2009) add that Millennials are strongly motivated by human values in action rather than by traditional incentives and status. Moreover, they increasingly apply a global mindset for solving societal problems and hence become a part of something larger than themselves. Pfeil (2014) concludes that German Millennials are mainly driven by the values loyalty, self-determination, freedom, universalism, and fairness. With reference to the workplace, the study reveals that a good work atmosphere, support for personal development, as well as congruent value schemes rank among the most important aspects for Millennials. Future expectations, employee loyalty and commitment are strongly value-driven among Millennials (Deloitte 2016). Deloitte's fifth global Millennial survey reveals that this generation has predominantly a positive view regarding the role of businesses in society but also expresses, that most businesses still insist on the profit motive alone. With regard to employee loyalty, 66% of Millennials expect to leave their organization by 2020. This lack of long-term commitment could be explained by the Millennial's perspective that only 28% of the organizations are making full use of the skills they currently have to offer. Although this generation has a generally positive view on the ability of businesses to do good, 64% state that businesses focus on their own agenda rather than considering wider society and 54% believe that businesses have no other ambitions beyond making money. Nevertheless, 58% of the respondents said that businesses behave in an ethical manner and 57% perceive leadership as committed to helping improve the society.

Millennials believe that the values which support the long-term success of a business are based on a fair and collaborative treatment of people, ethics, trust, integrity, and transparency and a strong customer focus. Moreover, a responsible behavior and its environmental impact are considered to be important factors as well. Only 5% see profit-focused values as an ensuring factor for long-term success. In addition, the survey's results outline that Millennials are strongly guided by their values in terms of choosing the workplace, fulfilling assignments, and making decisions. Contributing to a positive impact through the organizations they work at can be seen as a guiding principle. In addition, this cohort seems to be less impressed by the pure scale and prestige of business, rather than focusing on the positive contribution to the society at large (Deloitte 2016).

The connection between the satisfaction of essential expectations and the impact on their workplace loyalty is highlighted in Fig. 2.9 with reference to the aspects *Sense of Purpose* and *Aligned Values* (Deloitte 2016).

Considering the future workforce development, "Millennials will occupy every consequential leadership position in the world, be it in business, academia, government, or in the non-profit sector" (Universum 2015, p. 2). Hence, meeting their needs will be an ongoing task for corporations in terms of adapting their mindset towards this emerging workforce. In general, Millennials favor conscious work environments which reflect their values and support them in pursuing a meaningful

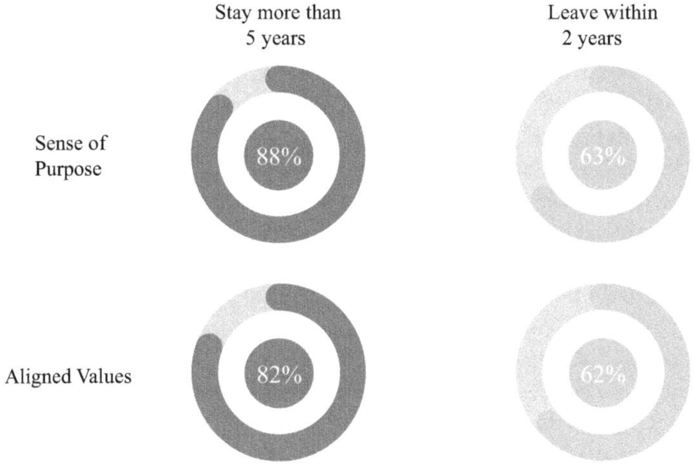

Fig. 2.9 Satisfaction and workplace loyalty among millennials (author's diagram)

work (Benett et al. 2009). One of their distinct characteristics in comparison to previous generations is their predominantly inherent access to digital technologies which represent a major aspect in nowadays information society (Goldman Sachs 2016).

2.6.3 Information in Society

Castells (2010) outlines that liberal and developed societies are shifting towards an information and knowledge society, which refers to an emergence of new social structures and networks. The linked diffusion of information transforms organizations in terms of structure and interaction with existing and potential customers.

Digital networking technologies tend to foster social and organizational networks in ways which allow their expansion and reconfiguration, overcome limitations of traditional networking forms of organization, and hence manage complexity beyond a certain size of the network. "The dynamics of nowadays information society are significantly shaping consciousness among people while factors such as traditions, family, religion, and traditional educational institutions tend to decrease" (Nefiodow and Nefiodow 2014, p. 209). Nefiodow and Nefiodow (2014) also see a shift regarding the flow of information from the fifth to the sixth Kondratieff cycle. Whereas the fifth cycle was characterized by a flow of information in and among operational systems and between human beings and machines, the sixth cycle is mainly driven by the flow of information in and among human beings as well as between human beings, the environment, and transcendence.

The internet shifted the balance of information power to the masses. It dramatically changed how people interact with each other, democratized information flows, and forced companies to operate with far greater transparency (Sisodia et al. 2014). From the year 2000 to 2014 the percentage of active internet users rose from 6.8 to 40.7% in 2014. In Germany, 86.2 out of 100 people are considered as internet users in 2014 (World Bank 2016b). Reputation is a crucial aspect for every organization and is likely to be damaged by company scandals, such as corruption, or other unethical behavior or practices. In contrast, reputation also represents a valuable source for long-term success in terms of customer attraction, employee retention, financial performance and overall stability (Eccles et al. 2007). The focus on reputation in nowadays information society has created an incentive to be open and transparent. This shift in mindset has impacted many businesses in towards a radical transparency policy and the way they conduct business as negative information are less likely to remain hidden (Thompson 2007). The claim for more transparency in organizations can be characterized as a crowd phenomenon. It is driven by both professional institutions as well as individuals who question the way a business is operating (Healy and Ramanna 2013). In addition, the emergence of whistleblowers can be identified as an additional factor which is mostly driven by a conflict of personal integrity and unethical company behavior (Venkatesan 2015). A study conducted by Klenk and Hoursch (2011) came to the conclusion that the perceived transparency of Dax-30 companies significantly correlates with the sympathy among internal and external stakeholders, and thus represents an essential aspect of trustworthiness. The shift in information power has given a competitive advantage for purpose-driven organizations which are adept at using their marketing tools not offensively to sell but to add value among the people who share their values and beliefs (Mackey and Sisodia 2014).

2.6.4 Conscious Consumer

Consumer decisions are based on an increasing amount of accessible information regarding products and services as well as the company behind it. Due to the partial loss of information control for corporations, a power shift can be observed in favor of consumers. Moreover, third party content such as reviews which is mainly created by consumers has become increasingly relevant in purchase decisions (Benett et al. 2009). Consumer generated virtual content through various platforms has digitalized word-of-mouth marketing and thus embodies an incremental shift in marketing practices and consumer empowerment. With regard to German consumers, 39% of all initial purchases were rated as being equally or more importantly influenced by digital word-of-mouth compared to traditional face-to-face recommendations.

In several industries, digital word-of-mouth marketing even outplaced traditional word-of-mouth marketing in terms of consumers' buying decisions. For example, digital sources accounted for more than 60% of all relevant purchase relevant information in case of the travel, entertainment electronics, and public utility services industry (Roland Berger 2014). Over the past 30 years, the number of conscious consumers has been steadily increasing. This consumer group believes that businesses should aim for good by creating value for everyone who is impacted (Honeyman 2014). Conscious consumers hold corporations morally responsible by putting pressure on ethically doubtful business decisions and consequently vote by their buying behavior (Aburdene 2005).

As Kip Tindell (2011), the CEO of the Container Store noted "everybody is becoming a little more conscious in the way they do business. I think customers demand it and they vote with their pockets". According to Benett et al. (2009) conscious consumers prefer products and services, which are offered by companies with a clear sense of human values and authenticity. The authors argue that this changing mindset is a result of mindless consumption, which caused undeniable damages to the environment as well as the society at large. Although environmental factors play a dominant role for conscious consumers, other aspects such as human labor conditions, compliance with laws, societal and philanthropic activities and a reflection of underlying human values in business guide their purchase decision as well. In order to simplify complex purchase decisions, conscious consumers seek for metrics such as carbon footprints, certifications, and other indicators, which contain information on the related impacts. Ultimately, this consumer group aims at democratizing purchase decisions by voting with their pocketbooks. In general, conscious consumption not only affects a certain type of offer but also the overall decreasing quantity of consumed resources (Benett et al. 2009). A study, conducted by the UK Office for National Statistics (2016) clearly indicated a constant decrease of raw material consumption from 2001 until 2013 in the UK and other European countries. This trend is expected to continue and is also supported by an increasing efficiency of material use. Nevertheless, conscious consumers still enjoy the pleasure of consumption combined with a positive contribution to the greater good. "The most successful corporate brands will be those that address these twin pulls of selfishness and altruism, and help consumers feel that what is good for them is also good for the world" (Benett et al. 2009, p. 31). Although some of these products and services might have an above-average pricing as a result of fair production conditions, conscious consumers are willing to pay a price premium (Network for Business Sustainability 2011). Conscious consumers are value and purpose-driven and aim at reflecting their beliefs in buying decisions. A recent study conducted by GlobeScan (2016a) indicates that there is a strong market for purpose-driven companies. Two out of three respondents stated that they aim at supporting companies that pursue a purpose of "making a positive difference in society through their products, services and operations" (GlobeScan 2016a, p. 2). Purpose has been perceived as an important element in building trust among customers and thus has direct implications regarding a company's success. Among the factors which embody a company's purpose, the products and services themselves were named as the most important information carrier, followed by company involvement in charities and communities, reputation and brand, ethical employee treatment, and efforts in environmental sustainability (GlobeScan 2016a). In terms of sustainable and responsible consumerism, a study among 6000 respondents from six countries clearly indicates that conscious consumption is necessary to preserve the environment for future generations (BBMG, GlobeScan, SustainAbility 2012).

Consumers from developing markets (China/India/Brazil) are twice as likely to purchase products because of their positive environmental and societal impact (51–22%) compared to consumers from developed markets (Germany/United Kingdom/United States). In terms of consumer segmentation, *Advocates* (14%) and *Aspirationals* (37%) represent the main supporters for conscious consumption, whereas *Practicals* and *Indifferents* rather orientate their buying-decision on a price-performance aspect (BBMG, GlobeScan, SustainAbility 2012).

Germany has been classified with 10% Advocates, 29% Aspirationals, 44% Practicals, and 10% Indifferents while China is divided in 6% Advocates, 53% Aspirationals, 31% Practicals, and 10% Indifferents. These results clearly indicate that conscious companies could aim at targeting their products for both markets and would potentially meet a strong acceptance in developing countries which match their customer profile (BBMG, GlobeScan, SustainAbility 2012). Especially, Aspirational consumers are expected to drive future consumer preferences and currently represent 40% of the global population (GlobeScan 2016b). However, the majority of Aspirationals have not established a map of purpose-driven brands. Hence, this circumstance represents a valuable opportunity for purpose-driven companies to attract and retain new customers. Around 63% of socially-conscious consumers are younger than 40 years old and thus are predominantly represented by Millennials and the Generation X (Nielsen 2012). Socially-conscious consumers predominantly rely on personal recommendations (95%), online consumer generated content and reviews (76%), and branded websites (65%). Among generations, Millennials can also be identified as the driving cohort behind the *sharing economy* (Goldman Sachs 2016). According to PwC (2015), 82% of the German Millennials already used an offer which is associated with the sharing economy. Although, Millennials represent the predominant cohort in this segment, it is expected the concept will increasingly gain acceptance among all age groups from 45% in 2015 to 64% in 2017. This development could be identified as an impact of the changing values from *owning* to *using* as a characteristic of the post-materialistic society.

2.7 Overview of Research Framework

This chapter describes the research framework (Fig. 2.10), which was applied for the following study. According to Mackey and Sisodia (2014) conscious businesses can be characterized by the four tenets *Higher purpose and core values*, *Conscious leadership*, *Conscious culture and management*, and *Stakeholder integration*. These parts represent the core aspects of an organizational implementation of conscious business principles and are aligned in an interrelated and reinforcing manner. The following chapters describe each of the four aspects and illustrate its practical implementation with a company example.

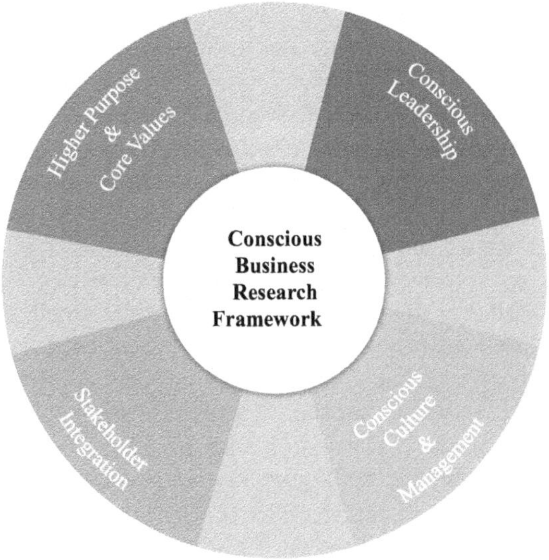

Fig. 2.10 Research framework—graphical overview (author's diagram)

2.7.1 *Higher Purpose*

According to Hurst (2016), the next economic era will be driven by the pursuit of personal, professional and organizational purpose which leads to a *Purpose Economy*. Conscious businesses focus on a purpose beyond profit (ConsciousCapitalism 2016). Every conscious business has defined a higher purpose, which addresses fundamental questions such as the business' reason for existence and its contribution to society (Mackey and Sisodia 2014). The higher purpose represents the core aspect which connects the other tenets and directs the implementation of a company's leadership, culture, and stakeholder orientation. Therefore, it represents the starting point in forming self-awareness, recognition, uniqueness, and its role for society. In addition, it serves as a communication instrument for internal and external stakeholders and embodies an essential part for aligning various interests towards a shared long-term goal. Focusing on purpose rather than profit is perceived as an incremental aspect to build confidence in business among its members and drive sustainable performance for the organization (Deloitte 2014). Experiencing a shared purpose leads from a disappearance of individual consciousness to a unifying group consciousness (Hsieh 2010). Hence, fostering a purpose-driven culture which is lived by its members, contributes to the organization's cohesion regarding a unified direction (Deloitte 2014). Thus, combining physical synchrony with other humans and being part of something bigger than oneself contributes to a greater sense of happiness (Hsieh 2010). Mourkogiannis et al. (2008) view the purpose of a company as superior to its tactics and strategies, as it embodies the most essential element in the constitution of a company and hence its moral DNA. Establishing a compelling higher purpose adds stability, decreases ambiguity in organizations, and enhances employee engagement and loyalty (Craig and Snook 2014). It embodies the set of reasons for conducting business, which resonates with people's ideas about what is right or worthwhile (Mourkogiannis 2007). Moreover, it regulates the dynamics of organizational behavior by aligning individual norms and beliefs towards a shared goal (Mourkogiannis et al. 2008). Furthermore, decision-making processes can be improved in terms of quality and time as this orientation provides the members of an organization with a common set of expected outcomes (Mackey and Sisodia 2014). Hence, a company's higher purpose has to be formulated in ways which can be translated into actual behavior and attitudes to fulfill its cohesive function and unite the organizations values. Otherwise, it is likely that a high variance of interpretations occur which dilute the common direction (Mourkogiannis et al. 2008). Purposeful organizational cultures have an inspiring, energizing and motivating characteristic. Moreover, purpose-driven motivation fosters *intrinsic motivation* and thus contributes directly to a high level of employee commitment and performance (Mackey and Sisodia 2014). Kofman (2006) adds that pursuing a noble purpose creates eudemonic happiness and fulfillment for those who actively participate in it. The experience of eudemonic happiness and fulfillment is likely to occur in situations where individuals perceive their work as meaningful towards others (Mourkogiannis et al. 2008).

Maintaining such an environment needs aligned members at different organizational levels and functions. Therefore, assessing a candidate's potential commitment to the company's higher purpose plays a crucial role for conscious businesses to keep this cohesion and make the right hiring decisions (Mackey and Sisodia 2014).

With reference to complementary concepts, Kenny (2014) points out that there is a clear distinction between a company's purpose, vision, and mission. "*Vision* refers to an imaginative conception of how the purpose can be achieved, while a company's *mission* represents the core strategy which is needed to realize its purpose" (Mackey and Sisodia 2014, 47). In addition, Mourkogiannis et al. (2008) point out that a company's purpose also represents a competitive factor as the market place. Market-based competition is not only shaped by products and services but also by moral ideals. Conscious businesses establish conditions, which energize and empower its members to align their personal passions with the company's higher purpose (Sisodia et al. 2014). When a company defines a social mission, employees feel better about their work and have a clearer sense of direction. Brands that are guided by a compelling mission, such as Patagonia and Timberland, can serve the best long-term, interests of both the brand and consumers (Armstrong et al. 2015). There is no intrinsic reason, why business should be different from any other individual human endeavor (Mackey and Sisodia 2014). With regard to this, Frankl (1976) notes that individuals should not search for an abstract meaning as everyone has his own purpose in life which is linked to a concrete assignment and demands fulfillment. This line of argument points out the importance of enduring ideals for individuals as well as organizations (Mackey and Sisodia 2014). These ideals were articulated by Plato as the transcendent of the Good, the True, and the Beautiful. The dimension of the Heroic nature has been added as many businesses pursue their purpose driven by this motive (Mackey and

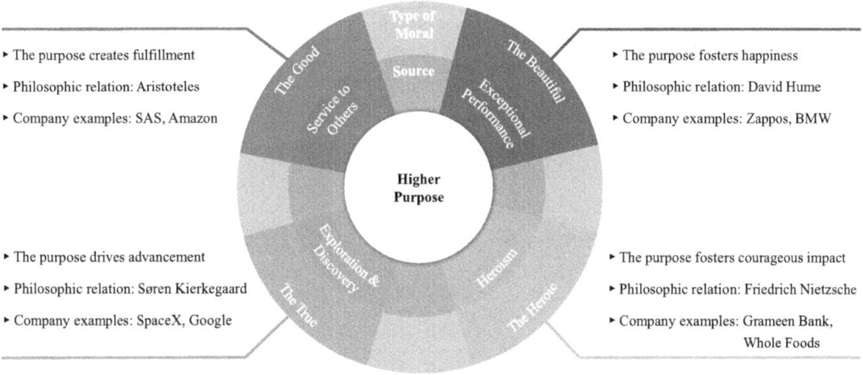

Fig. 2.11 Categories of higher purposes (author's diagram)

Sisodia 2014). Figure 2.11 combines the findings of Mourkogiannis et al. 2008 as well as Mackey and Sisodia 2014.

The Good embodies a deeply motivating purpose that is emotionally fulfilling for individuals who truly embrace this ideal. Authentic service is based on genuine empathy with the needs and desires of others. Pursuing this ideal is closely linked to an increased level of love, care, and compassion with regard to the customer relationship. Companies with a clear emphasis on service to others strongly depend on the goodwill of their customers. Therefore, customer-centric industries such as retail place a strong emphasis on aligning their business accordingly. Examples for this alignment are the SAS, The Container Store, Amazon, Nordstrom, Starbucks and other (Mackey and Sisodia 2014).

The True is driven by a search for truth and the pursuit of knowledge. Companies that fall into this category, see their purpose closely linked to the fostering of human advancement, increasing the quality of life, and accelerating human knowledge. Examples of this category are Google, Intel, Genentech and SpaceX.

The Beautiful can be expressed through a pursuit of beauty, excellence, the quest for perfection and performance. Companies like Apple are driven to create products, which are aligned to this dimension in terms of design and functionality. A further example for this ideal is BMW.

The Heroic characterizes an ideal, which is mainly driven by a desire to change the world. The will to contribute to a better world translates into a strong service to others or through discovery and the pursuit of truth, through the quest for perfection, and solving extensive problems. The Grameen Bank, which was established by Mohammad Yunus in Bangladesh, can be named as an example for this dimension. His will to end poverty and uplift underprivileged people by supporting them with micro credits has been widely accepted as a heroic business. Whole Foods can be named as another representative of this category (Mackey and Sisodia 2014). The EY Beacon Institute conducted a survey among 474 executives in cooperation with the Harvard Business Analytics Services regarding the impact

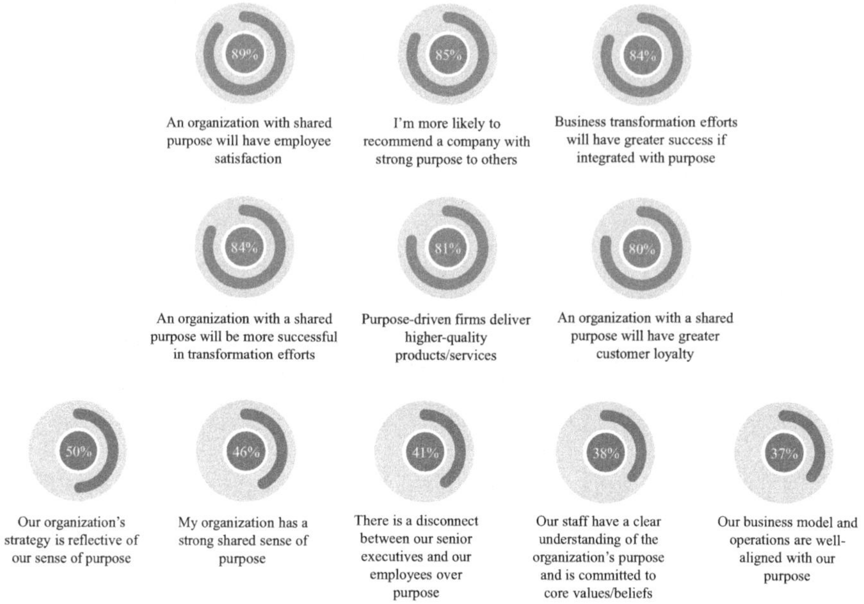

Fig. 2.12 Purpose at the workplace (author's diagram)

of purpose at the workplace (HBRAS 2015). A variety of implications and insights is outlined within Fig. 2.12.

Figure 2.13 shows how the dimension *The Good* translates into business practices at Zappos.

2.7.2 Stakeholder Integration

Mackey and Sisodia (2014) view stakeholders in general as the entities that impact or are impacted by a business (Table 2.3). Stakeholder relationships are based on legally explicit contracts as well as emotional, mostly implicit contracts. Both contract forms are the base for building trust and a lasting connection between different parties in a business context. Forming a positive emotional bond with stakeholders is an essential aspect regarding a reciprocal exchange of trust and alignment. Conscious businesses appreciate the nature of both contracts and act accordingly to their responsibility in an ethical and mutually valuable way (Sisodia et al. 2014). All impacted entities contribute to the creation of value, and it is therefore vital that they share fairly in the distribution of the created value. Conscious businesses endear themselves to their stakeholders by aligning the interests in a synergetic way and thus reject the idea of favoring one stakeholder group at the expense of another. Freeman et al. (2007) add that the rejection of trade-off thinking has to be done consistently in order to act with integrity and

Company Profile

Zappos.com Inc.		
	Founded	1999 in San Francisco (USA)
	Industry	Retail
	Headquarter	Las Vegas
	CEO	Tony Hsieh
	# Employees 2016	~ 1300
	Revenue 2016	~ $2 billion

The retail company Zappos was founded in 1999 in San Francisco by Nick Swinmurn and Tony Hsieh, who still leads the company as CEO. The vision of offering the best customer service online has been a founding principle for the company. In 2009, Amazon acquired Zappos for $1.2 billion. This transaction was mainly successful because of a shared passion for customer service and the vision to leverage mutual benefits for both companies. Since its foundation, Zappos has been steadily increasing their product portfolio from shoes to other categories such as fashion, watches and other (Zappos 2016a).

The Zappos culture has always been a distinct success factor for the company. Zappos core values are expressed by the following ten aspects (Zappos 2016b).

- Deliver WOW Through Service
- Embrace and Drive Change
- Create Fun and A Little Weirdness
- Be Adventurous, Creative, and Open-Minded
- Pursue Growth and Learning
- Build Open and Honest Relationships with Communication
- Build a Positive Team and Family Spirit
- Do More with Less
- Be Passionate and Determined
- Be Humble

In his book *Delivering Happiness: A Path to Profits, Passion, and Purpose,* Tony Hsieh outlines his blueprint regarding the connection between purpose and happiness. The higher purpose to deliver happiness for various stakeholders and especially customers is a dominant explanation for the success of Zappos. According to Hsieh (2010), the establishment of a higher purpose enables a long-term fulfillment and happiness for the entire organization

Hsieh (2010) emphasizes that there is an essential connection between time and the level of happiness. Thus, *Pleasure* represents a hedonic state which lasts only for a short period of time. The next state is characterized as *Passion* which incorporates flow and engagement. The highest form of happiness is titled *Higher Purpose and Meaning* and translates into a perception of being part of something bigger than yourself.

Establishing an organization, which is directed by a higher noble purpose leads to significant higher levels of happiness and positively contributes to the overall success. Companies with a higher sense of purpose perform up to four times better compared to those without (Hsieh 2010). Zappos has developed different measures to foster its higher purpose on the dimensions, *community, employees,* and *the world* (Zappos 2016c).

In terms of linking the company purpose to the community, Zappos employees engage in charity events such as *Bald & Blue, The Great Turkey Giveaway, Roos-N-More* and others. The proceeds of these events benefit different communitarian initiatives and involve purposeful behavior by employees (Zappos 2016c).

Increasing employee happiness by organizing new hire *Graduation Celebrations,* free *Health and Wellness* treatments, and other company events further strengthen its purpose (Zappos 2016c). Moreover, a company *culture book,* an open library, internal blog, and other instruments are designed to connect employees to the purpose of delivering happiness (Hsieh 2010). Delivering happiness represents the main objective at Zappos to align stakeholder interests and performance.

Fig. 2.13 Purpose implementation at Zappos (author's diagram)

Table 2.3 Stakeholder matrix (author's table)

Stakeholder	Description
Society	Local and broader communities, governmental and non-governmental institutions, future generations
Partners	Upstream partners such as suppliers, horizontal partners, and downstream partners such as retailers
Institutions	Governmental and non-governmental institutions as well as related administrative and political representatives
Investors	Individual and institutional shareholders as well as creditors
Customers	Individual and organizational customers
Employees	Current, future, and past employees and their families
Media	Institutionalized and social media channels
Environment	Local and global environment
Competitors	Direct and indirect competitors

create trust among various interest groups. This perspective shift marks a key characteristic of all conscious businesses compared to traditional business models. Managing stakeholders from this point of view requires an understanding of the interconnectedness and interdependent relationships among the parties involved. Hence, creating value for all is an important and complex aim and requires a high level of systemic intelligence. Coping with the complexity of stakeholder management represents a success factor for every organization and therefore requires a systemic and holistic view to mitigate risks and create value for all impacted entities (Mackey and Sisodia 2014). Considering various stakeholder interests requires a high level of network intelligence for individuals and organizations (Friedman and Miles 2006). Hence, being aware of manifest and potential stakeholder needs forms a common ground for growth. Therefore, conscious businesses foster a mindset of collaboration, co-creation, and interdependency to create a basis of long-term harmonious business relationships (Schwerin 2011). Organizations which are dominated by a traditional *either/or* attitude will fail to cope with an increasing level of stakeholder complexity as this mindset is inadequate for discovering new opportunities as a result of interdependent interests. Hence, conscious businesses, which apply a *both/and* perspective can harness and manifest potential opportunities by making use of their stakeholder network and connect partners to create mutual value (Sisodia et al. 2014).

Carroll and Buchholtz (2015) refer to the classification of stakeholder groups with regard to an integral approach. Although, these stakeholders are organized in an integral approach, conscious businesses set priorities regarding each group (Mackey and Sisodia 2014). Peter Drucker (2008b, p. 98) states that the only valid definition of business purpose is "to create a customer". With reference to this, Mackey and Sisodia (2014) note that most conscious businesses either place their customers or employees on the highest priority in terms of value creation. Whole Foods Market sees their customers as the most important stakeholder as they build the basis for sustaining their business model in a competitive environment.

Hence, forming a close emotional connection to customers and valuing their individual needs can be considered as primary goals in forming a high level of customer loyalty and satisfaction. Acting with integrity, transparency, love, respect, and authenticity and treating them as part of the Whole Foods Market family is a prerequisite to developing a high-trust relationship. Managing one stakeholder on the expense of another reflects a traditional business mindset. Therefore, conscious businesses focus less on *who* should be treated with the highest priority than *what* should be addressed with a higher priority and urgency (Mackey and Sisodia 2014). This orientation is reflected by the company's higher purpose and therefore offers guidance towards balancing stakeholder priorities (Sisodia et al. 2014). However, creating trustful and caring relationships is an inherent principle regarding all stakeholders. SAS Institute has been focusing on developing their high-trust culture among various interest groups. As a result, the company could integrate their internal and external stakeholders as an aligned coalition which supports their mission (Hurley 2012). Sisodia et al. (2014) outline that living and promoting a culture of trust and ethical behavior towards internal and external stakeholders is a key prerequisite for every promotion decision among SAS employees. Moreover, the CEO of SAS, Jim Goodnight (Goodnight and Floriada 2005) sees it as an absolute necessity, to empower employees to make use of their creative potential and thus help the company to succeed. As every company relies on the acceptance of the communities in which it operates, society can be classified as a fundamental interest group among all stakeholders.

Therefore, conscious businesses focus on serving communities to help facilitate their public purpose (Sisodia et al. 2014). Based on this understanding, Whole Foods Market founded the Whole Planet Foundation in 2005 with the mission to alleviate poverty through microcredit in communities around the world that supply Whole Foods Market stores with products. The donation-based initiative already supported more than 7.8 million people in 68 countries with committed funds of almost $69 million and a repayment rate of 97% (Whole Planet Foundation 2016). Every business consumes scarce natural resources, and therefore has to find ways on how to compensate and foster the preservation for future generations. The environment is the only silent stakeholder and therefore has to be closely considered by every business to uncover negative effects, which are caused by its operations (Mackey and Sisodia 2014). Creating consciousness among individual as well as organizational impacts on nature is critical in managing environmental risks and ultimately maintain a biosphere for all living beings (Hawken et al. 1999). The B Corporation, Seventh Generation is a producer and distributer of environmentally safe household products, paper products, and personal care. The company strongly focuses on the conservation of natural resources by reusing materials for production and plant-based ingredients for in their products (Seventh Generation 2016a). Seventh Generation is driven by the mission to inspire a consumer revolution that nurtures the health of the next seven generations. The company releases every year a *Corporate Consciousness Report* to set the course for a more mindful way of doing business and inform internal and external stakeholders about their progress in doing so (Seventh Generation 2016b). In 2015, 70% of their products and packaging created zero waste, were biodegradable or recyclable and 81% of the used ingredients and materials were bio-based or recycled. The aspiration for 2020 is to achieve 100% for all aspects and control the necessary steps in a transparent way for their stakeholders. Their accomplishments have been awarded by the B Corp Environmental Impact Award, the Carbon Cup, and further honor (Seventh Generation 2016c). The Corporate Consciousness Manager, Ashley Orgain (2015) concludes that integrating stakeholders in a synergetic way is the base for the company to achieve its long-term goals to change the world. Every company is in need of capital to develop its business and pursue its purpose. Businesses have an ethical responsibility to respectfully use the capital which they receive by investors to create value for both sides (Mackey and Sisodia 2014). The goal of conscious companies is to achieve their higher purpose in a profitable way (Kofman 2013b). With regard to this Kofman (2013a) outlines that strong profits will not be sustainable without establishing an equally strong interpersonal solidarity and personal well-being among all stakeholders. As investors play an important role among every company's stakeholder network, conscious businesses treat them equally respectfully as internal customers. A main distinguishing factor between investors of traditional and conscious businesses is reflected by their usually different investment horizon. Conscious businesses do not try to attract short-term investors and speculators as they aim at creating long-term value rather than short-term wins. Hence, conscious businesses try to attract investors who believe in their higher purpose based on a high level of trust, transparency, integrity, patience, and support (Mackey and Sisodia 2014). Figure 2.14 shows the integral stakeholder concept of Whole Foods Market, which represents the underlying base for its value delivery among all stakeholder groups.

2.7.3 Conscious Culture and Management

"For individuals, character is destiny. For organizations, culture is destiny" (Hsieh 2010, p. 210). Schein (1992, p. 12) defines culture as a "pattern of basic assumptions—invented, discovered, or developed by a given group as it learns to cope with problems of external adaption and internal integration—that has worked well enough to be considered valid and, therefore, to be taught to new members as the correct way to perceive, think, and feel in relation to those problems." It is both a "here and now" dynamic phenomenon and builds a structure which influences its members in multiple ways (Schein 2010, p. 3). An organization's culture is constantly reenacted by the connections and interactions of the organization's members and thus shaped by their values, norms, beliefs, and behavior (Schein 2010). Moreover, it represents the psychosocial infrastructure, which directs individual behavior through a pattern of shared similarities and goals (Sisodia et al. 2014). In terms of an organization's effectiveness, Schein (2010) views culture as a limiting and determining factor, which builds the basis for an effective and successful collaboration. "Developing a conscious culture is a business imperative.

Company Profile

Whole Foods Market Inc.

Founded	1980 in Austin (USA)
Industry	Food
Headquarter	Austin
CEO	John Mackey & Walter Robb
# Employees 2016	90,900
Revenue 2016	$15.724 billion

Whole Foods Market was founded in 1980 with the purpose to co-create a world which enables individuals, communities, and the planet to flourish (Whole Foods 2016a). The four founding members partnered to create the first supermarket format for natural and organic food combined with a strong commitment to sustainable agriculture (Whole Foods 2016b).

John Mackey, one of the founders and co-author of the book Conscious Capitalism established an integral concept of aligning different stakeholder interests. This concept neglects zero-sum thinking which implies that stakeholder interests can only be managed by trade-offs rather than a mutual base for benefit (Mackey and Sisodia 2014).

Whole Foods has established a *Declaration of Interdependence*, which promotes their perspective of interdependence among all stakeholders by transparent communication and active integration (Whole Foods 2016c). Nevertheless, Whole Foods has a clear prioritization concerning their primary interest groups. Customers are put first, followed by employees, suppliers, and lastly investors.

The society as a whole, the environment and the government are perceived as essential surrounding stakeholders. To enable active collaboration with their primary stakeholders, Whole Foods established an initiative called Future Search. Representatives of each group define the company's vision for the next five years in a mutual agreement and discuss ways to include their different interests. For instance, employees in each market are empowered to decide for their team benefits such as special health plans, retirement packages and others. Whole Foods established this policy to democratize their decision making process and thus value individual perspectives (Mackey and Sisodia 2014).

Compassionate listening to stakeholder interests and collaboration with integrity embodies the basis of the Whole Foods integral stakeholder management approach. Supporting communities worldwide through the Planet Foundation or regular employee and supplier engagement in communitarian and environmental initiatives are further examples of their approach which is illustrated below according to Mackey (2006).

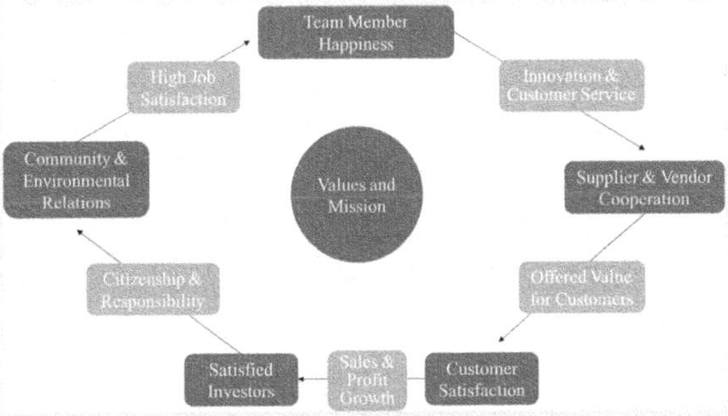

Fig. 2.14 Whole foods stakeholder integration model (author's diagram)

Culture undergirds an organization. It enables the execution of the organization's strategy, the achievement of its goals, and the fulfillment of its mission" (Kofman 2006, p. 15). A conscious business' culture represents a source of strength and stability for its members and ensures that its purpose and core values endure over time and through potential changes in leadership. Conscious cultures naturally evolve from the company's commitment to pursue its higher purpose, foster stakeholder interdependence, integrate its stakeholders, and lead consciously. As a prerequisite for sustaining a conscious culture, shared values, norms and beliefs have to be consistent with the ones from the environments, in which they operate to reflect a common mindset. Conscious cultures are characterized by their self-sustaining, self-healing, and evolutionary nature. Moreover, they are enduring and resilient towards potentially harmful external influences. Aligning interdependent stakeholders in this structure requires a high level of collaboration, which has to be embodied by each member of a conscious organization. Hence, collaboration builds the fundamental element on how individuals behave internally and externally (Sisodia et al. 2014). The collaborative benefit can be described as the *social capital* of an organization and has to be recognized as such by leaders and managers (Sivachithappa, 2014). Aligning individual energy to an unified group goal represents an integral task for every organization. An organization's energy can be characterized by its intensity and quality. The amount of activity, interaction, alertness, and emotional arousal represent the intensity, whereas the quality is influenced by its positive or negative emotional states. High levels on both dimensions results in a so called *passion zone* which energizes the entire organization and thus represents a desired zone in to foster creativity, innovation, and performance (Bruch and Ghoshal 2003). Conscious businesses aim at fostering a positive and collaborative culture among their members and stakeholders, which is based on humanity, compassion, and happiness (Fryer 2013). As a result, this orientation is also likely to translate into higher levels of employee productivity, customer loyalty, and financial performance (Sisodia et al. 2014). Mackey and Sisodia (2014) summarize seven qualities of conscious cultures as trust, accountability, caring, transparency, integrity, loyalty, and egalitarianism (TACTILE), which are clarified in Fig. 2.15.

Enabling conscious businesses to leverage from the outlined seven qualities requires a virtuous managerial approach which aims at authentic inspiration of others. Thus establishing conscious management requires the organization to support the development of self-management as well as the management of others (Ganguli 2011). Conscious businesses view their members as valuable individuals, who are supported in their personal aims and aspirations. As a result, greater creativity and innovation are likely to occur compared to transactional management approaches. "If management views workers not as valuable, unique individuals but as tools to be discarded when no longer needed, then employees will also regard the firm as nothing more than a machine for issuing paychecks, with no other value or meaning" (Csikszentmihalyi 2003, p. 101). With regard to this, an organizational mindset of collaboration is essential as innovation without a culture of collaboration results in egocentric innovation and will unlikely benefit the company's collective

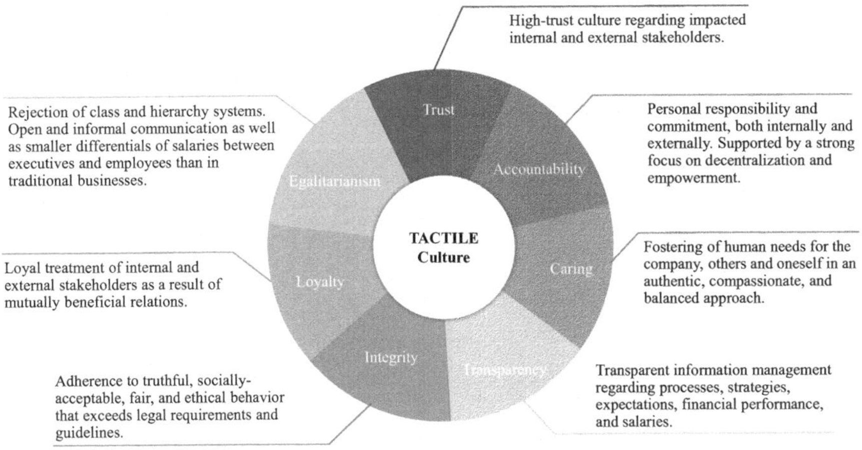

Fig. 2.15 Seven qualities of conscious cultures (author's diagram)

intelligence. A distinct characteristic of collaborative and high-trust conscious cultures can be observed by a strong focus on self-organization in order to achieve a higher level of agility, accountability, integrity, and empowerment. Empowering employees by democratizing decision-making processes creates a new level for opportunities and innovation, and supports the conscious development of organizations. Applying principles of holocracy gives conscious businesses the opportunity to leverage their full potential (McGrath 2013). The Brazilian centrifuge manufacturer, Semco Partners can be outlined as one company example which successfully applies the principles of a democratized and self-organizing business (Gebhard 2016). Zappos represents another conscious business which is organized according to Holocracy and self-organization. Tony Hsieh (2016) favors this organizational structure as it allows employees to act as entrepreneurs who benefit from an increased self-direction. Another example embodies the German IT Solution provider Velian. The company aims to fully integrate holocracy as part of their conscious business activities by the end of 2016 (Velian 2016). "Conscious managers exercise a minimal amount of control. Their role is not to control other people; it is to create the conditions that allow for more self-management" (Mackey and Sisodia 2014, p. 249).

In terms of an organization's cultural consciousness, Barrett (2017a) developed a model which illustrates the spectrum of consciousness and its relation to various stakeholders. The model is structurally based on the hierarchy of needs by Abraham Maslow and is divided into seven levels. Barrett (2017b) notes that conscious businesses have to fulfill needs on all levels in order to be stable, sustainable, and beneficial for all impacted stakeholders. Each stage relates to a specific existential need that is common to the human condition and thus represents the principal motivating drivers. The model can be applied to all forms of organizations regardless of their industry focus or profit motive. Figure 2.16 focuses on profit businesses.

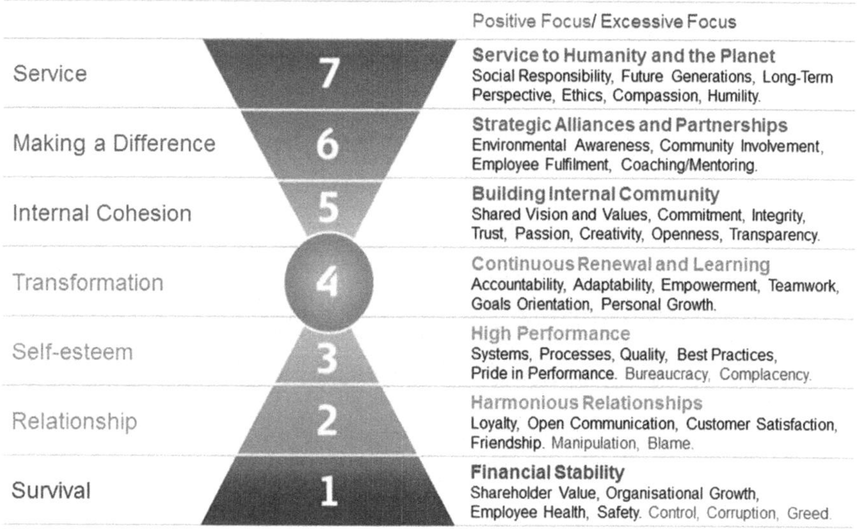

Fig. 2.16 Levels of organizational consciousness (Barrett 2017a)

The first stage, survival focuses on the pursuit of profit and shareholder value and is driven by the motive of stability. The second stage, relationship puts an emphasis on building harmonious relationships that create a sense of belonging and loyalty among employees as well as establishing a close bond between the organization and its customers. Self-esteem relates to high performance systems and processes that foster pride among the organization's members. This phase mainly targets a reduction of traits that hinder performance and order. The transformation stage focuses on adaptability and continuous learning by empowering employees in an environment that highly values innovation, continuous improvement, knowledge sharing as well as the growth of each individual. Internal cohesion is characterized by a strong cohesive culture that enhances the organization's capacity for collective action. This direction is achieved by aligning individual motivation towards a unified mission and vision which is complemented by a shared set of values with regard to a collective commitment, integrity, enthusiasm, creativity and passion. The sixth's stage *making a difference* focuses on strategic alliances and partnerships to advance in global challenges such as environmental protection. At the same time, organizations on this level show a high degree of internal connectivity by fostering collaboration across business units and departments. The final stage *service* relates to the social responsibility by engaging with stakeholders and other organizations in pursuit of societal objectives which enhance the sustainability of humanity and the planet. Moreover, this stage embodies a deepened level of internal connectivity and collaboration based on compassion, humility and forgiveness.

The described stages portray different levels of organizational needs that have implications regarding the needs of its stakeholders. The first level primarily targets investors and employees in contrast to a stronger orientation towards customers and employees on the second stage. The levels three to five put an emphasis on employee needs which is complemented by partners and local communities in the sixth's stage. Finally, the seventh stage also addresses the needs of the society as a whole.

With reference to Maslow's hierarchy of needs, the first three stages can be described as *basic* business needs which are driven by the organization's and the investor's self-centeredness. These *deficiency* needs do not embody a source of long-term satisfaction as the leader's fear to miss these needs usually create fear-driven organizations that are prone to external influences. The transformational phase marks a shift from fear-driven to purpose and value driven organizations. This translates into an orientation towards adaptive system that shows a higher level of resilience, employee empowerment and accountability.

In accordance to the Maslow's terminology, the stages five to seven represent *growth* needs of organizations. These needs bundle cultural cohesion, synergetic alliances and partnerships as well as long-term sustainability and social responsibility. If these needs are constantly met and cultivated, organizations predominantly show a high degree of lasting commitment and motivation.

Organizations that only focus on *basic* business needs are caught in self-centeredness and are unable to adapt in an agile way to environmental changes due to their excessive focus on hierarchy and bureaucracy. Creativity and innovation usually fall short as a result of a fear-driven market orientation rather than actively driving change, development and advancement. Moreover, employees usually show low levels of commitment, enthusiasm and creativity which translate into low satisfaction and high rates of turnover and absenteeism. On the other hand, an exclusive focus on *higher* needs without taking the *basic needs* into account will result in a non-effective and instable organization. This instability is rooted in missing structures, procedures and financial management that are necessary for a healthy business system. Hence, organizations need to combine *basic* and *growth* needs to operate as full-spectrum conscious businesses.

Southwest Airlines serves as an example of a conscious business which includes a high degree of the outlined spectrum and is summarized in Fig. 2.17.

2.7.4 Conscious Leadership

"Leadership is the process by which a person sets a purpose for other persons and motivates them to pursue it with effectiveness and full commitment. Leadership transforms individual potential into collective performance" (Kofman 2006, p. 9). Aligning conscious employees towards a unifying higher purpose requires conscious managers and leaders. Aligning individual interest towards a tribe mindset builds the base for transforming personal interests towards a shared goal (Logan et al. 2011). In order to form and lead a tribe, leaders have to establish a sense of connectedness among the people, who work together and embrace shared values and happiness (Hsieh 2010). Conscious leaders act as servants who subordinate

Company Profile

Southwest Airlines Co.

Founded	1967 in Dallas (USA)
Industry	Aviation
Headquarter	Dallas
CEO	Gary C. Kelly
# Employees 2016	53,072
Revenue 2016	$15.35 billion

Southwest Airlines was established in 1967 by Herb Kelleher and Rollin King. Starting with a budget of $500,000, the airline emerged in-to the world's largest low-cost carrier with 97 destinations in eight different countries. The airline has become the largest carrier of domestic passengers in the United States and employs more than 49,000 people. Southwest Airlines is publicly traded under the symbol LUV and has demonstrated a profitable financial performance over the last decades (Southwest Airlines 2016a).

One of the essential success factors of Southwest Airlines is represented by its corporate culture and leadership which was mainly driven by its founder and former CEO, Herb Kelleher (Kochan 1999). The company is guided by its higher purpose to connect people to what is important in their lives through friendly, reliable, and low-cost air travel (Southwest Airlines 2016b). Southwest Airlines transforms this purpose into the vision of becoming the world's most loved, most flown, and most profitable airline (Southwest Airlines 2016c). Realizing this purpose and vision is based on the perspective to value employees as the single greatest strength and most enduring long-term competitive advantage (Kelly 2016).

The company emphasizes the importance of its core values *Servant's heart* and *fun-loving attitude* to create a conscious culture. In addition, the airline focuses on their *Warrior Spirit* which encompasses performance-oriented aspects as well as their unique work in their *Southwest Way attitude*, which emphasizes their purpose to offer friendliness, reliability, and cost efficiency (Southwest Airlines 2016c).

Southwest Airlines applies a people-centric view and set the priority on the welfare for employees, customers, and the communities in which they operate. A key element in establishing a constantly high customer loyalty level represents the focus on employee happiness and engagement (Dunn 2015). Creating a positive, inspiring, and compassionate work atmosphere has been the primary goal for the company (Sisodia et al. 2014). This direction was manly driven by the former CEO and co-founder Herb Kelleher and translated into their company-wide Culture Committee (Dunn 2015).

This committee was formed in 1990, and consists of changing members of the organization who represent all regions and departments among the Southwest system (Southwest Airlines 2016d). The responsibility of this group is to do whatever is necessary to create, enhance, and enrich the special Southwest spirit and culture that has made it to a impactful company and family" (Heskett 2003). The committee consists of people who embody inspiring storytellers and are able to act as the company's cultural ambassadors, and missionaries. The committee members are selected according to their demonstrated level of creativity, performance, expertise, energy, enthusiasm, and most importantly their representation of the Southwest spirit (Freiberg and Freiberg 1998). The members are responsible for helping to communicate important company information from senior leaders, brainstorming new company initiatives, and volunteering for culture events designed to engage and celebrate employees. Currently, the committee consists of 157 active members who are selected for a three-year term (Dunn 2015).

Fig. 2.17 Conscious culture at Southwest Airlines (author's diagram)

individual interests and ego-centric focus towards the pursuit of a common good (Secretan 2001). This connection is a prerequisite for enabling full organizational performance and effectiveness (Kofman 2006). Hence, conscious leadership plays the most important role in transforming purpose into action and empowering every member of an organization to contribute with their full potential. Conscious leaders are required to integrate a philosophical alignment, which centers humanity as an inherent element of their actions and attitudes. Therefore, leaders in conscious organizations embody the driving force to shape a business which integrates and fosters wisdom and compassion in support of human development (Kofman 2006). With reference to this, Bill George (2014) remarks that conscious leadership is synonymous to the concept of authentic leadership. Authentic leaders perform their work with superior personal integrity and are committed to build enduring organizations, which embody a deep sense of purpose and are true to their core values. Moreover, authentic leaders orientate their decisions towards a synergetic benefit among all impacted stakeholders and recognize their service to society (George 2015). Conscious leadership embodies the link between ancient wisdom and modern economic effectiveness in an integral way (Kofman 2006). Conscious leaders are primarily motivated by achieving the firm's higher purpose and the creation of value for all stakeholders. This leadership concept focuses on a *both/and* mindset and thus rejects a zero-sum, trade-off perspective, where one stakeholder can only benefit on the expense of another (Mackey and Sisodia 2014). Conscious leaders recognize the unique values of the others involved, and seek to amplify them for a collective synergetic good (Kaipa and Radjou 2013). They lead by example and focus on developing, inspiring, and empowering their employees to pursue the company's higher purpose collectively. Conscious leaders incorporate feminine values such as caring, compassion, and cooperation as well as rather masculine values such as assertiveness, competition, and ambition and are able to use both aspects with regard to the context. Moreover, these leaders have commonly established a high level of analytical, emotional, spiritual, and systems intelligence. Leading employees and partners with empathy request a high level of emotional control, to manage interests and potential conflicts as emotions define the possible range of behavior. Fear and other negative emotions hinder the ability to make conscious decisions as they decrease the range of possible actions (Kofman 2006). The involved level of consciousness determines the impact on transforming dysfunctional behavior and unconscious mental pattern among involved peers (Anderson and Anderson 2001). Leading others from a positive perspective in cases of conflict and uncertainty is therefore a quality, which has to be developed by every conscious leader.

This development can be supported by reflecting on self-awareness, self-acceptance, self-regulation, self-enquiry, and self-inquiry in a mindful and compassionate manner (Kofman 2006). Mindful leadership refers to the ability to be conscious of changes, which occur in internal and external contexts without judging or trying to control them. This approach offers a welcoming notion towards change and transformation as well as seeing the bigger picture and deciding for an adequate response by using a holistic perspective (Kaipa and Radjou 2013). Therefore, connecting with employees to support the collective pursuit of the company's higher purpose requires leaders to cultivate their emotional intelligence. "Emotional intelligence is the capacity for recognizing our own feelings and those of others, for motivating ourselves, for managing emotions well in ourselves and in our relationships" (Goleman 1998, p. 317). This form of intelligence reflects the ability to read and understand others in social contexts and utilizes such knowledge to compassionately influence others through emotional regulation and control (Prati et al. 2003). Emotional intelligence combines intrapersonal and interpersonal intelligence and knowledge through empathy, compassion, and self-awareness. Among those qualities, self-awareness marks the fundamental ability of emotional intelligence, as it regulates the understanding of oneself and others. Nevertheless, this quality has been widely ignored in many organizations regarding their leadership development (Goleman 2002). More than 1500 CEOs advocated in a worldwide survey that leadership styles need to change in order to cope with an increasing level of complexity in business (IBM 2010). Cherniss and Goleman (2001) note that developing emotional intelligence among leaders represents an essential part in enabling organizations to understand complex business environments and integrate stakeholder interests accordingly. Hence, Voola et al. (2004) consider emotional intelligence as a competitive advantage and a necessity for strategic change and transformation. The increasing level of complexity in business has forced organizations to transform their perspective on how to conduct business. The perspective of viewing organizations as rational hierarchical machines has widely vanished towards a new understanding, which perceives businesses as complex organisms which require a systemic management approach (Prati et al. 2003). Gantt and Agazarian (2004) outline that a systems-centered perspective requires individuals to understand organizations as interconnected entities in a dynamic internal and external environment. Hence, businesses need to adapt their collective systems intelligence to reflect on their impact and manage possible actions and reactions on all levels (Kaipa and Radjou 2013). Fostering collective systems intelligence is driven by conscious leaders, who have an inherent systemic sensitivity. Moreover, they are able to transform their knowledge into action and share their wisdom with employees to enable a collaborative learning process. Conscious leaders promote the advancement of the organization as a whole rather than focusing on local sub-systems. Optimizing sub-systems can put the organization at large to an existential risk as these systems are usually aligned in a conflicting structure and thus do not contribute to a common goal (Kofman 2010). Therefore, leaders with well-developed systems intelligence direct their actions on the premise to assess systemic impacts and synergetic interests (Secretan 2001). System-centric leaders are able to anticipate immediate and long-term consequences of actions understand the root of organizational behavior and consciously shape the corporate architecture (Mackey and Sisodia 2014).

Spiritual Intelligence refers to the "ability to act with wisdom and compassion while maintaining inner and outer peace (equanimity), regardless of the circumstances" (Wigglesworth 2004, p. 2). It enables individuals to access their deepest meanings, values, purposes, and highest motivations. Moreover, spiritual intelligence encompasses the innate ability to distinguish right from wrong and thus builds the base for developing goodness, truth, beauty, and compassion in our lives (Zohar and Marshall 2004). "Spiritual does not have a primarily religious connotation but refers to the specifically human dimension" (Frankl 1976, p. 160). Kofman (2006) states that there is no tension between material and spiritual wealth when business is conducted with a high level of consciousness. Conscious leaders view their work as an expression of spiritual practice and thus do it for a cause far greater than themselves (Brown 2013). The Global Centre for Conscious Leadership (2016) developed an overview regarding the complex web of conscious leadership and its interrelated connections which is shown in Fig. 2.18.

Brown (2013) concludes that conscious leaders need to combine horizontal and vertical learning processes in order to lead successfully. Horizontal learning encompasses the development of competences, such as functional knowledge, skills and behaviors. Vertical learning refers to the transformation of a leader's mindset and represents all aspects on how individuals learn and interpret situations. Hence, this type of learning refers to the aspects of emotional, systemic, and spiritual intelligence. However, traditional leadership development concepts have been mainly focusing on the horizontal dimension (Brown 2013). Rooke and Torbert (2005) developed a model for assessing vertical learning in business environments. Barrett (2017c) adapted Maslow's hierarchy of needs and formed a classification on different levels of conscious leadership as outlined in Fig. 2.19.

According to Barrett (2017d), *Crisis Manager* embodies the first stage of leadership consciousness and focus the security of basic organizational needs and stability in adverse business environments. The next level is defined as *Relationship Manager* and relates to an open communication to recognize individual needs and solve conflicts. In addition, this role emphasizes the creation of employee and customer satisfaction and dignified leadership style. In order to foster organizational self-esteem, the third stage addresses high performance and process excellence in an *organizing* manner. Strategic planning and efficiency orientation build the ground for performance related pride. Transforming an organization requires a *facilitating* and *influencing* stage of leadership consciousness that focuses on empowerment and continuous learning within the organization. This mindset and behavior is a necessary prerequisite to ensure individual growth and spark innovation though collaborative teamwork, courage, initiative and accountability. Fostering internal cohesion is reflected by an *integrating* and *inspiring* level of consciousness that creates a strong cohesive culture and offers capacity for collective action towards a unified goal. Purpose, vision, mission and values are at the center and accompanied by authenticity, creativity, integrity and passion. The sixth's stage embodies a level that is focused on a *mentor* and *partner* role to cultivate strategic alliances and partnerships in a servant leadership manner. Furthermore, this level addresses employee fulfillment though empathetic mentoring and coaching and intuition based decision making processes. Advancing service to society, humanity and the environment is achieved by a *wisdom* and *visionary* understanding of leadership consciousness. This direction translates into a strong

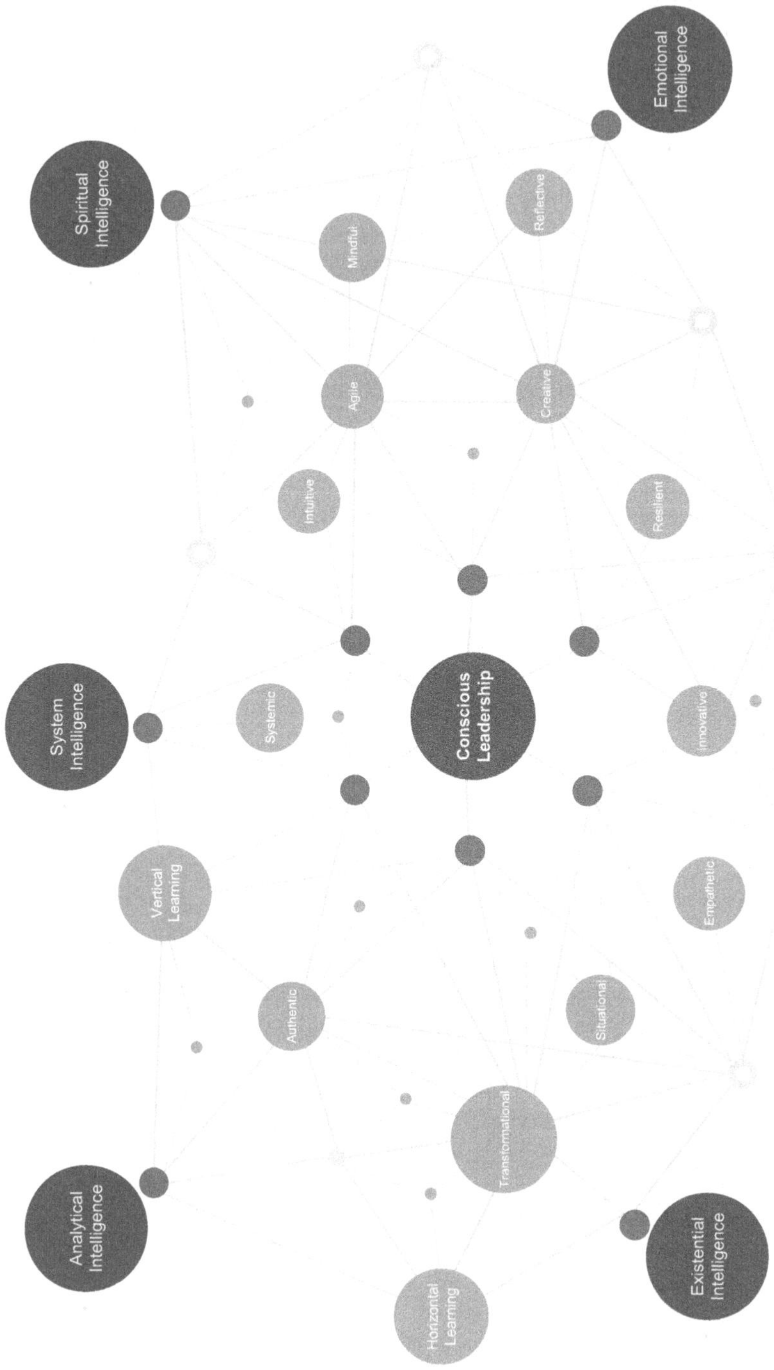

Fig. 2.18 Conscious leadership system (author's diagram)

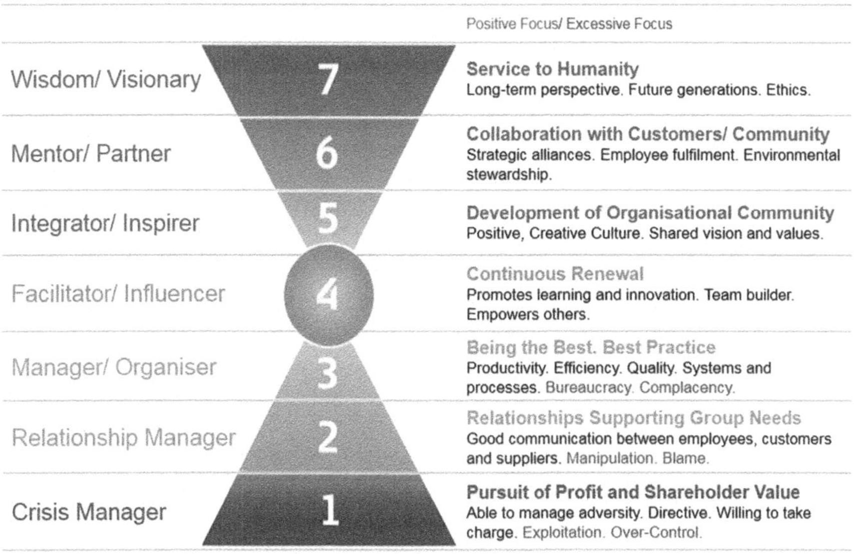

Fig. 2.19 Levels of leadership consciousness (Barrett 2017c)

emphasis on ethics, social responsibility, sustainability and the well-being of future generations by applying wisdom, compassion and humility.

The first three stages can be a significant source for negative traits as unhealthy behavior mainly derives from fear-driven attitudes. These individual fears can translate into a variety of perceived deficiencies related to power, recognition, acceptance and other. With regard to this background, Barrett (2017e) describes this state the *personal entropy* as the amount of fear-driven energy that a person expresses in his or her day-to-day interactions to ensure the functioning of a system. Thus, leadership styles can be classified due to their levels of entropy and therefore the effectiveness to lead in a value-adding manner. Hence, becoming a conscious leader demands self-mastery concerning fear-driven motivation and leading in purpose-driven way. In addition, conscious leaders need to incorporate a high degree of self-reflection to be aware of their driving forces in any given situation to assess from which mental source they are deriving their decisions. Excessive control, confusion, internal competition, silo-mentality, strict hierarchy and other traits can cause a high degree of organizational dysfunction. Lee (2010) and Barrett (2010) emphasize that there is a direct link between personal entropy of leaders and the well-being of organizations as the level of personal entropy directly influences cultural entropy in organizations. This form of entropy expresses the amount of energy that is related to deficient work and thus a measure of potential conflict, friction and internal frustration. Chappell and Deweby (2014) remark that entropy in organizations encompasses horizontal level of disorder within tiers (e.g. same department) and vertical entropy that measures the disorder along lineages or between tiers (e.g. different business units). Moreover, it has direct implications towards employee satisfaction and commitment. The amount of value-added energy that employees are able to contribute to the organization is equal to their input energy minus the energy that is required to ensure the functioning of the organization minus the cultural entropy. Overall, a high level of cultural entropy reduces the energy available for value-added work and thus results in low organizational performance and efficiency (Barrett 2010).

Brown (2013) notes that highly conscious leaders are represented by level six and seven. Leaders on level five are characterized as *from Good to Great* and levels one to four are synonymous with conventional leaders who are driven by a rather traditional leadership approach. Developing conscious leaders has a high priority at LinkedIn in order to achieve the company's purpose. Figure 2.20 summarizes key aspects and the underlying mindset which is applied at LinkedIn.

2.8 Hypotheses

This paragraph outlines three main study hypotheses that represent an underlining basis for the case study and the correlation to findings from the theoretical framework. The hypotheses will be evaluated on a qualitative approach as this study does not include a statistical exploration. Mackey and Sisodia (2014) state that conscious businesses are characterized by pursuing a purpose beyond profit by applying a humanistic mindset, an integrative stakeholder management concept as well as a conscious approach to leadership, management and culture. As a result, conscious businesses are able to benefit from this orientation on a long-term perspective. This also relates to an above-average financial performance which is indicated by Sisodia et al. (2014). The findings are based on conscious businesses which are predominantly based in the United States. Therefore, this study aims at assessing conscious business principles and their impact on performance in German companies.

H1: Conscious business practices have been contributing to a long-term sustainable growth model and a high degree of organizational performance for conscious businesses in Germany.

As conscious business represents a concept which became manifest in the early 2000s, it is subject to question which factors can be identified as key drivers. Delbanco (1999), Sisodia et al. (2014) Bakan (2004), and Nefiodow and Nefiodow (2014) name a value shift in Western societies and the constant growth of information as distinct drivers. Hence the second hypothesis is formulated as follows.

H2: Consciousness in business is predominantly driven by a demographic value change and a constant growth and accessibility of information.

The B Corporations as well as the Economy of the Common Good include various characteristics of conscious business and represent an new movement in contrast to traditional businesses. Both initiatives have been expanding their

Company Profile

LinkedIn Inc.

Founded	2002 in Mountain View (USA)
Industry	Online Platform
Headquarter	Mountain View
CEO	Jeff Weiner
# Employees 2016	~ 10,000
Revenue 2016	~ $3 billion

LinkedIn represents the world's largest professional network with more than 433 million members in 200 countries (LinkedIn 2016a). It is based on the vision to create economic opportunity for every member of the global workforce by connecting the world's professionals to make them more productive and successful (Weiner 2016). In June 2016, Microsoft announced to acquire LinkedIn to leverage and benefit from both business models by combining a professional network with professional cloud solutions. Both CEOs state that their companies are run by a similar purpose and structure which will lead to synergies (Weiner 2016).

Establishing an open and compassionate leadership style at LinkedIn has been an essential part of their development (Weiner 2012). Maintaining and fostering a compassionate work culture is supported by offering tailored coaching as well as mindfulness meditation classes. LinkedIn is focused on putting their values into practice. Managers and leaders at LinkedIn are assessed regarding their impact in terms of fostering compassion and wisdom among employees (Weiner 2012). Their culture of transformation is key for consciously evolving, oneself, the company, and consequently the world (LinkedIn 2016b).

In November 2013, Fred Kofman the author of Conscious Business joined LinkedIn as Philosopher and Vice President. His main area of work combines leadership and organizational development within the company. One of his main contributions to the organization has been the creation of the Conscious Business Academy. This academy focuses on the principles of conscious business and aims at introducing and teaching the core elements to the community (Conscious Business Academy 2016). In addition to the Conscious Business Academy, LinkedIn established the Conscious Business Leadership Awards, which embody a platform to recognize and honor efforts of conscious leaders in the Asia-Pacific region (LinkedIn 2016c).

The Conscious Business Academy offers a certification program which is aligned to the conscious business principles by Fred Kofman. The program was first designed to foster a culture of conscious transformation at LinkedIn, and is now available for free to the public. The content is mainly focused on leadership and management topics for individuals as well as for the entire organization (Conscious Business Academy 2016). The program consists of the following twelve chapters (LinkedIn 2016c).

- Introduction
- Leadership
- Culture
- Response-Ability
- Humility
- Communication
- Collaboration
- Coordination
- Co-Evaluation
- Perspectives
- Emotions
- Mindfulness

Fig. 2.20 Conscious leadership at LinkedIn (author's diagram)

activities globally and grew in terms of its members. Two out of the selected companies have become a part of this movement in the last years. Hence, the third hypothesis assumes that this development will continue and will challenge traditional businesses towards a new way of conducting business.

H3: Conscious businesses in Germany have increased in importance and hence will challenge traditional business practices towards a new paradigm in the future.

References

Abelshauser W (2011) Deutsche Wirtschaftsgeschichte: Von 1945 bis zur Gegenwart. C.H. Beck, München

Aburdene P (2005) Megatrends 2010: the rise of conscious capitalism. Hampton Roads, Charlottesville, VA

Allianz Global Investors (2010) The sixth kondratieff—long waves of prosperity. https://www.allianz.com/v_1339501901000/media/press/document/other/kondratieff_en.pdf. Accessed 15 June 2016

Anderson D, Anderson LA (2001) Beyond change management: advanced strategies for today's transformational leaders. Pfeiffer, San Francisco, CA

Armstrong G, Adam S, Denize S, Kotler P (2015) Principles of marketing. Pearson, Melbourne

B Lab (2016a) The B Corp declaration. https://www.bcorporation.net/what-are-b-corps/the-b-corp-declaration. Accessed 11 June 2016

B Lab (2016b) Performance requirements. https://www.bcorporation.net/become-a-b-corp/how-to-become-a-b-corp/performance-requirements. Accessed 11 June 2016

B Lab (2016c) Welcome. https://www.bcorporation.net/. Accessed 8 June 2016

Bakan J (2004) The corporation: the pathological pursuit of profit and power. Free, New York

Baptist EE, Hyman L (2014) American capitalism: a reader. Simon & Schuster, New York

Barrett R (2010) High performance—it's all about entropy. https://www.valuescentre.com/sites/default/files/uploads/2011-12-14/High%20Performance%20-20It%27s%20all%20about%20entropy.pdf. Accessed 15 Jan 2017

Barrett R (2017a) Levels of organisational consciousness. https://www.valuescentre.com/mapping-values/barrett-model/organisational-consciousness. Accessed 10 Jan 2017

Barrett R (2017b) The seven levels of organisational consciousness. https://www.valuescentre.com/sites/default/files/uploads/2010-07-06/The%207%20Levels%20of%20Organisational%20Consciousness.pdf. Accessed 10 Apr 2017

Barrett R (2017c) Levels of leadership consciousness. https://www.valuescentre.com/mapping-values/barrett-model/leadership-consciousness. Accessed 12 Jan 2017

Barrett R (2017d) The seven levels of leadership consciousness. https://www.valuescentre.com/sites/default/files/uploads/the_7_levels_of_leadership_consciousness.pdf. Accessed 14 Apr 2017

Barrett R (2017e) Personal entropy. https://www.valuescentre.com/mapping-values/leadership/personal-entropy. Accessed 12 May 2017

Baumann H, Schulz S (2015) Allgemeine Bevölkerungsumfrage der Sozialwissenschften: ALLBUS 2014 Variable Report. https://dbk.gesis.org/DBKSearch/download.asp?id=56577. Accessed 21 June 2016

BBMG, GlobeScan, SustainAbility (2012) Re: thinking consumption: consumers and the future of sustainability. http://www.globescan.com/component/edocman/?task=document.viewdoc&id=51&Itemid=0. Accessed 2 July 2016

Benett A, Gobhai C, O'Reilly A, Welch G (2009) Good for business: the rise of the conscious corporation. Palgrave Macmillian, New York

Bennis W (2014) Foreword to the first edition. In: Sisodia R, Sheth J, Wolfe D (eds) Firms of endearment: how world-class companies profit from passion and purpose, 3rd edn. Pearson Education, New York, pp x–xiv

Berman B (2011) Competing in tough times: business lessons from L.L. Bean, Trader Joe's, Costco, and other world-class retailers. Pearson, Upper Saddle River, NJ

Bertelsmann Stiftung (2015) Bedeutung der Arbeit: Ein Kooperationsprojekt von GfK Verein und Bertelsmann Stiftung. https://www.bertelsmann-stiftung.de/fileadmin/files/BSt/Publikationen/GrauePublikationen/Bedeutung_der_Arbeit_final_160114.pdf. Accessed 8 July 2016

Branden N (1999) The art of living consciously: the power of awareness to transform everyday life. Fireside, New York

Brandt F, Georgiou K (2016) Shareholders vs. stakeholders. http://scholarship.law.upenn.edu/cgi/viewcontent.cgi?article=1002&context=fisch_2016. Accessed 22 June 2016

Brown BC (2013) The future of leadership for conscious capitalism. https://associates.metaintegral.org/sites/default/files/MetaIntegral_Brown_The%20future%20of%20leadership%20for%20conscious%20capitalism_20131018.pdf. Accessed 21 June 2016

Bruch H, Ghoshal S (2003) Unleashing organizational power. http://sloanreview.mit.edu/article/unleashing-organizational-energy/. Accessed 24 Mar 2016

Carroll AB, Buchholtz AK (2015) Business and society: ethics, sustainability, and stakeholder management. Cengage, Stamford, CT

Castells M (2010) The rise of the network society. Wiley-Blackwell, Chichester

Chappell D, Deweby GT (2014) Defining the entropy of hierarchical organizations. Complex Govern Netw 1(2):41–56. https://doi.org/10.7564/14-CGN17

Cherniss G, Goleman D (2001) Emotional intelligence and organization effectiveness, the emotional intelligent workplace: how to select for, measure, and improve emotional intelligence in individuals, groups, and organizations. Jossey-Bass, San Francisco, CA

Collins J (2001) Good to great: why some companies make a leap and other's don't. HarperCollins, New York

Conscious Business Academy (2016) Content. https://conscious.linkedin.com/content. Accessed 13 May 2016

ConsciousCapitalism (2016) Higher Purpose. http://www.consciouscapitalism.org/purpose. Accessed 11 June 2016

Corfe R (2008) Social capitalism in theory and practice: emergence of the new majority. Arena, Bury St. Edmunds

Costco (2015) Annual report 2015. http://phx.corporate-ir.net/phoenix.zhtml?c=83830&p=irol-irhome&cm_re=1_en-_-Bottom_Nav-_-Bottom_investor&lang=en-US. Accessed 13 June 2016

Craig N, Snook SA (2014) From purpose to impact. Harv Bus Rev 92(5):104–111

Csikszentmihalyi M (2003) Good business, leadership, flow, and the making of meaning. Viking, New York

Delbanco A (1999) The real American dream: a mediation on hope. Harvard University Press, Cambridge

Deloitte (2014) Culture of purpose—building business confidence; driving growth: 2014 core beliefs & culture survey. http://www2.deloitte.com/content/dam/Deloitte/us/Documents/about-deloitte/us-leadership-2014-core-beliefs-culture-survey-040414.pdf. Accessed 1 July 2016

Deloitte (2016) The 2016 Deloitte millennial survey winning over the next generation of leaders. https://www2.deloitte.com/content/dam/Deloitte/global/Documents/About-Deloitte/gx-millenial-survey-2016-exec-summary.pdf. Accessed 3 July 2016

Drucker P (2008a) The essential Drucker: the best of sixty years of Peter Drucker's essential writings on management. Harper Collins, New York

Drucker P (2008b) Management: revised edition. HarperCollins, New York

Dunn C (2015) Southwest: a triple bottom line airline. http://www.consciouscompanymagazine.com/blogs/press/37973953-southwest-a-triple-bottom-line-airline. Accessed 15 June 2016

Eccles RG, Newquist SC, Schatz R (2007) Reputation and its risks. Harv Bus Rev 85(2):104–114
Economy for the Common Good (2016) The content of the balance sheet. https://www.ecogood.org/en/balance-sheet/what-balance-sheet/contents-balance-sheet. Accessed 8 June 2016
EESC (2015) Stellungnahme des Europäischen Wirtschafts- und Sozialausschusses zum Thema "Die Gemeinwohl-Ökonomie: Ein nachhaltiges Wirtschaftsmodell für den sozialen Zusammenhalt" (Initiativstellungnahme). http://eur-lex.europa.eu/legalcontent/DE/TXT/?uri=CELEX%3A52015IE2060. Accessed 2 June 2016
European Commission (2011) Corporate social responsibility: a new definition, a new agenda for action. http://europa.eu/rapid/press-release_MEMO-11-730_en.htm. Accessed 12 June 2016
Felber C (2012) Die Gemeinwohl-Ökonomie: Eine demokratische Alternative wächst. Deuticke, Wien
Felber C (2014) Die Gemeinwohl-Ökonomie: Eine demokratische Alternative wächst. Deuticke, Wien
Felber C (2015) Europäische Union diskutiert Modell der Gemeinwohl-Ökonomie. https://www.ecogood.org/allgemeine-infos/aktuelles/neuigkeiten/europaeische-union-diskutiert-modell-der-gemeinwohl-oekonomie. Accessed 3 June 2016
Financial Times (2015) FT 500 2015. http://im.ft-static.com/content/images/b38c350e-169d-11e5-b07f-00144feabdc0.xls. Accessed 18 June 2016
Forsa (2014) Gesund im Job. https://www.dak.de/dakonline/live/dak/download/ForsaUmfrage_zum_Thema_Gesund_im_Job-1471144.pdf?. Accessed 20 June 2016
Frankl VE (1976) Man's search for meaning. Pocket, New York
Freeman ER (2002) The possibility of stakeholder capitalism. In: Zsolnai L, Gasparski WW (eds) Ethics and the future of capitalism: paxiology: the international annual of practical philosophy and methodology, vol 9. Transaction, New Brunswick, NJ, pp 111–116
Freeman ER, Harrison JS, Wicks AC (2007) Managing for stakeholders: survival, reputation, and success. Yale University Press, New Haven, CT
Freiberg K, Freiberg J (1998) Nuts: Southwest Airlines crazy recipe for business and personal success. Broadway, New York
Friedman M (1970) The social responsibility of business is to increase its profits. http://query.nytimes.com/mem/archivefree/pdf?res=9E05E0DA153CE531A15750C1A96F9C946190D6CF. Accessed 11 June 2016
Friedman AL, Miles S (2006) Stakeholders: theory and practice. Oxford University Press, Oxford
Fryer B (2013) The rise of compassionate management (finally). https://hbr.org/2013/09/the-rise-of-compassionate-management-finally/. Accessed 21 June 2016
Gallup (2016) Millennial banking customers: two myths, one fact. http://www.gallup.com/businessjournal/187364/millennial-banking-customers-two-myths-one-fact.aspx. Accessed 18 June 2016
Ganguli S (2011) Conscious management: managing and leading happily at the workplace and at home. Allied, New Delhi
Gantt SP, Agazarian YM (2004) Systems-centered emotional intelligence: beyond individual systems to organizational systems. Organ Anal 12(2):147–169. https://doi.org/10.1108/eb028990
Gates B (2008) A new approach to capitalism. In: Kinsley M (ed) Creative capitalism: a conversation with Bill Gates, Warren Buffet, and other economic leaders. Simon & Schuster, New York, pp 7–17
Gebhard C (2016) The definitive guide to conscious business. http://www.cipcg.com/downloads/The%20Definitiv%20Guide%20To%20Conscious%20Business.pdf. Accessed 18 June 2016
Gemeinwohl-Ökonomie (2016a) Geschichte der Gemeinwohlökonomie. https://www.ecogood.org/allgemeine-infos/idee/geschichte-der-gemeinwohl-oekonomie. Accessed 6 June 2016
Gemeinwohl-Ökonomie (2016b) Ecogood-Map. https://www.ecogood.org/allgemeine-infos/aktuelles/neuigkeiten/ecogood-map-die-interaktive-landkarte-der-gwoe-ist-online. Accessed 3 June 2016

Gemeinwohl-Ökonomie (2016c) Inhalte der Gemeinwohlbilanz. https://www.ecogood.org/gemeinwohl-bilanz/was-ist-die-gw-bilanz/inhalte-der-gemeinwohl-bilanz. Accessed 3 June 2016
George B (2014) Getting capitalism back on track. In: Mackey J, Sisodia R (eds) Conscious capitalism: liberating the heroic spirit of business. Harvard Business Review, Boston, MA, pp ix–xiv
George B (2015) Authentic leadership rediscovered. http://hbswk.hbs.edu/item/authentic-leadership-rediscovered. Accessed 22 June 2016
Gexsi (2016) Die B Corp Zertifizierung auf einem Blick. http://goodbusiness.news/die-b-corp-zertifizierung-auf-einem-blick/. Accessed 8 June 2016
GfK (2016) Hauptsache, sicher: Umfrage des GfK Vereins zur Bedeutung von Werten. http://www.gfk-verein.org/sites/default/files/medien/359/dokumente/pm_gfk_verein_bedeutung_von_werten_2015_2016_fin.pdf. Accessed 3 July 2016
Global Centre for Conscious Leadership (2016) Conscious leadership blueprint. http://gcfcl.com/conscious-leadership-blueprint/. Accessed 18 June 2016
GlobeScan (2009) Wide dissatisfaction with capitalism—twenty years after fall of berlin wall. http://www.globescan.com/news_archives/bbc2009_berlin_wall/bbc09_berlin_wall_release.pdf. Accessed 10 June 2016
GlobeScan (2016a) The public on purpose: executive summary: insights from a global study on corporate purpose. http://www.globescan.com/component/edocman/#download-modal-content. Accessed 12 July 2016
GlobeScan (2016b) Aspirational consumers are rising. Are brands ready to meet them?. http://www.globescan.com/news-and-analysis/press-releases/press-releases-2016/390-aspirational-consumers-are-rising-are-brands-ready-to-meet-them.html. Accessed 3 July 2016
Goldman Sachs (2016) Millennial: coming of age. http://www.goldmansachs.com/our-thinking/pages/millennials/. Accessed 28 June 2016
Goleman D (1998) Working with emotional intelligence. Bantam, New York
Goleman D (2002) Primal leadership, with Daniel Goleman. http://iveybusinessjournal.com/publication/primal-leadership-with-daniel-goleman/. Accessed 24 Mar 2016
Goodnight J, Floriada R (2005) Managing for creativity. Harv Bus Rev 83(7):124–131
Goyal A, Goyal M (2009) Business environment. V.K., New Delhi
Great Place to Work (2016) The container store. http://reviews.greatplacetowork.com/the-container-store. Accessed 17 June 2016
Gupta CB (2007) Business studies. Tata McGraw-Hill, New Delhi
Hawken P, Lovins A, Lovins HL (1999) Natural capitalism. Brown, Boston, MA
HBRAS (2015) The business case for purpose. http://www.ey.com/Publication/vwLUAssets/ey-the-business-case-for-purpose/$FILE/ey-the-business-case-for-purpose.pdf. Accessed 12 Feb 2017
Healy PM, Ramanna K (2013) When the crowd fights corruption. Harv Bus Rev 91(1):122–128
Heskett JL (2003) Southwest Airlines 2002: an industry under siege. https://hbr.org/product/southwest-airlines-2002-an-industry-under-siege/an/803133-PDF-ENG. Accessed 12 June 2016
Honeyman R (2014) The B Corp handbook: how to use business as a force for good. Berrett-Koehler, San Francisco, CA
Hsieh T (2010) Delivering happiness: a path to profits, passion, and purpose. Business Plus, New York
Hsieh T (2016) Holocracy and self-organization. https://www.zapposinsights.com/about/holacracy. Accessed 18 June 2016
Hurley RF (2012) The decision to trust: how leaders create high-trust organizations. Jossey-Bass, San Francisco, CA
Hurst A (2016) The purpose economy, expanded and updated: how your desire for impact, personal growth and community is changing the world. Elevate, Boise

IBM (2010) Capitalizing on complexity insights from the global chief executive officer study. http://www-01.ibm.com/common/ssi/cgi-bin/ssialias?htmlfid=GBE03297USEN&appname=wwwsearch. Accessed 3 June 2016
Idowu S, Kasum A, Mermod A (2014) People, planet, and profit: socio-economic perspectives of CSR. Gower, Burlington, VT
Kaipa P, Radjou N (2013) From smart to wise: acting and leading with purpose. Jossey-Bass, San Francisco, CA
Kaku R (1997) The path of Kyosey. Harv Bus Rev 74(4):55–63
Kaplan (2016) MCAT behavioral sciences review: comprehensive review for all MCAT behavioral sciences topics. Kaplan, New York
Kelly G (2016) About Southwest. https://www.southwest.com/html/about-southwest/index.html?int=. Accessed 16 June 2016
Kenny G (2014) Your company's purpose is not its vision, mission, or values. https://hbr.org/2014/09/your-companys-purpose-is-not-its-vision-mission-or-values. Accessed 10 June 2016
Kinsley M (2008) Creative capitalism: a conversation with Bill Gates, Warren Buffet, and other economic leaders. Simon & Schuster, New York
Klenk & Hoursch (2011) Transparenz-Studie 2011 Berichtsband. http://www.transparenz.net/wp-content/uploads/2011/07/Berichtsband_Transparenz-Studie.pdf. Accessed 8 July 2016
Koch C (2012) Consciousness: confessions of a romantic reductionist. MIT Press, Cambridge, MA
Kofman F (2006) Conscious business: how to build value through values. Sounds True, Boulder, CO
Kofman F (2010) Conscious business: how to build value through values. http://www.axialent.com/pdf/Conscious%20Business%20by%20Fred%20Kofman%20-%20Ch%201%20English.pdf. Accessed 3 June 2016
Kofman F (2013a) Conscious business: how to build value through values. Sounds True, Boulder, CO
Kofman F (2013b) LinkedIn speaker series: Fred Kofman. https://www.youtube.com/watch?v=IdMvWLARF1w. Accessed 12 June 2016
Kromphardt J (2015) Analysen und Leitbilder des Kapitalismus von Adam Smith bis zum Finanzkapitalismus. Metropolis, Marburg
Lachman ME (2001) Handbook of midlife development. Wiley, New York
Lawrence AT, Weber J (2011) Business and society: stakeholders, ethics, public policy. McGraw-Hill, New York
Lee H (2010) Leadership success factors—examining leadership qualities by personal entropy. https://www.valuescentre.com/sites/default/files/wysiwyg-images/Leadership_Success_Factors.pdf. Accessed 14 Mar 2017
Lerner RM, Easterbrooks MA, Mistry J (2003) Handbook of psychology, Volume 6, Developmental psychology. Wiley, Hoboken, NJ
LinkedIn (2016a) About us. https://www.linkedin.com/about-us. Accessed 13 June 2016
LinkedIn (2016b) LinkedIn's culture of transformation. https://blog.linkedin.com/2015/03/11/linkedins-culture-of-transformation. Accessed 13 June 2016
LinkedIn (2016c) Content. https://conscious.linkedin.com/content. Accessed 14 June 2016
Logan D, King J, Fischer-Wright H (2011) Tribal leadership: leveraging natural groups to build a thriving organization. Harper Business, New York
Mackey J (2006) Conscious capitalism: creating a new paradigm for business. http://www.wholefoodsmarket.com/blog/john-mackeys-blog/conscious-capitalism-creating-new-paradigm-for%C2%A0business. Accessed 20 June 2016
Mackey J (2016) An introduction to conscious capitalism. Conscious capitalism exists to elevate humanity. http://www.consciouscapitalism.org/node/3998. Accessed 2 June 2016
Mackey J, Sisodia R (2014) Conscious capitalism: liberating the heroic spirit of business. Harvard Business School Publishing, Boston, MA
McGrath A (2013) Holocracy: creating conscious organizations. http://wonderworksconsulting.com/wp-content/uploads/2013/10/Creating_Conscious_Org.pdf. Accessed 18 June 2016

Mourkogiannis N (2007) On purpose. Bus Strateg Rev 18(1):38–41. https://doi.org/10.1111/j.1467-8616.2007.00452.x
Mourkogiannis N, Vogelsang G, Unger S (2008) Der Auftrag: Was großartige Unternehmen antreibt. Wiley-VCH, Weinheim
Müller H (2012) Wertewandel. http://www.bpb.de/politik/grundfragen/deutsche-verhaeltnisse-einesozialkunde/138454/werte-milieus-und-lebensstile-wertewandel. Accessed 8 June 2016
Nefiodow L, Nefiodow S (2014) The sixth kondratieff: a new long wave in the global economy. Rhein-Sieg, Sankt-Augustin
Network for Business Sustainability (2011) Socially conscious consumerism. http://nbs.net/wp-content/uploads/NBS-Consumerism-Primer.pdf. Accessed 20 June 2016
Nielsen (2012) The global, socially-conscious consumer. http://www.nielsen.com/content/dam/corporate/us/en/reports-downloads/2012-Reports/Nielsen-Global-Social-Responsibility-Report- March-2012.pdf. Accessed 4 July 2016
OECD (2013) Public spending on health and long-term care: a new set of projections. https://www.oecd.org/eco/growth/Health%20FINAL.pdf. Accessed 22 June 2016
Opaschowski HW (2010) Grundlagenstudie zum Wertewandel in Deutschland: Das Ende der Ichlinge. http://www.stiftungfuerzukunftsfragen.de/fileadmin/_migrated/media/Forschung-Aktuell-226-Ende-der-Ichlinge.pdf. Accessed 4 July 2016
Orgain A (2015) Transform commerce. 2015 corporate consciousness report. http://www.seventhgeneration.com/sites/default/files/2015_seventh_generation_corporate_responsibility_report.pdf. Accessed 20 June 2016
Orizon (2014) Studie Arbeitsmarkt 2014: Perspektive der Arbeitnehmer—Zentrale Ergebnisse. https://www.orizon.de/uploads/tx_ozttnews/Orizon_Diagramme_Arbeitsmarktstudie_2014_gesamt_15122014_01.pdf. Accessed 3 July 2016
Patagonia (2011) Don't buy this jacket. http://www.patagonia.com/email/11/112811.html. Accessed 17 June 2016
Patagonia (2016a) Patagonia's mission statement. http://www.patagonia.com/us/patagonia.go?assetid=2047. Accessed 17 June 2016
Patagonia (2016b) Better than new: worn wear. http://www.patagonia.com/us/worn-wear. Accessed 17 June 2016
Patagonia (2016c) Worn wear. http://eu.patagonia.com/deLU/worn-wear. Accessed 3 June 2016
Pew Research Center (2016) Millennials overtake baby boomers as America's largest generation. http://www.pewresearch.org/fact-tank/2016/04/25/millennials-overtake-baby-boomers/. Accessed 20 June 2016
Pfeil S (2014) Enactus-Studie 2014 Das Arbeitgeberwahlverhalten der Generation Y: Eine werteorientierte Analyse unter besonderer Berücksichtigung des Sinns der Arbeit. http://www.upj.de/fileadmin/user_upload/MAIN-dateien/Infopool/Forschung/enactus_gysummary_2014.pdf. Accessed 8 June 2016
Polman P (2015) Paul Polman on multinationals, public markets, and the B Corp movement. https://www.youtube.com/watch?v=yJQhkHCJkqE. Accessed 8 May 2016
Poon B (2010) TEDxToronto—Bruce Poon Tip "beyond the triple bottom line". https://www.youtube.com/watch?v=xbeUftMoFUQ. Accessed 12 June 2016
Porter ME, Kramer MR (2002) The competitive advantage of corporate philanthropy. Harv Bus Rev 90(12):1–15
Porter ME, Kramer MR (2006) Strategy and society: the link between competitive advantage and corporate social responsibility. Harv Bus Rev 84(12):78–92
Prati ML, Douglas C, Ferris GR, Ammeter AP, Buckley RM (2003) Emotional intelligence, leadership effectiveness, and team outcomes. Int J Organ Anal 11(1):21–40. https://doi.org/10.1108/eb028961
PwC (2015) Share Economy: Repräsentative Bevölkerungsbefragung. https://www.pwc.de/de/digitale-transformation/assets/pwc-bevoelkerungsbefragung-share-economy.pdf. Accessed 5 July 2016

Renesch J (2012) The conscious organization: workplace for the self-actualized. Spanda J 3 (1):227–230

Roland Berger (2014) German digitalization consumer report. https://www.rolandberger.de/media/pdf/Roland_Berger_German_Digitalization_Consumer_Report_20140718.pdf. Accessed 1 July 2016

Rooke D, Torbert WR (2005) Seven transformations of leadership. Harv Bus Rev 83(4):66–76

Rosser JB, Rosser MV (2004) Comparative economics in a transforming world economy. MIT Press, Cambridge, MA

Sanders T (2008) Love is the killer app: how to win business and influence friends. Three Rivers, New York

Schein EH (1992) Organizational culture and leadership, 2nd edn. Jossey-Bass, San Francisco, CA

Schein EH (2010) Organizational culture and leadership, 4th edn. Jossey-Bass, San Francisco, CA

Schröder M (2014) Varianten des Kapitalismus: Unterschiede liberaler und koordinierter Marktwirtschaften. Springer, Wiesbaden

Schwab K (2016) Legitimationsprobleme im kaltlaufenden Kapitalismus: Für soziale Unternehmer. http://www.nzz.ch/meinung/debatte/fuer-soziale-unternehmer-1.18716549. Accessed 2 July 2016

Schwerin D (2011) Conscious capitalism: principles for prosperity. Routledge, New York

Secretan L (2001) The conscious leader. Industry Week 250(2):19–20

Senge P, Scharmer OC, Jaworski J, Flowes BS (2004) Presence: human purpose and the field of the future. SoL, Cambridge, MA

Seventh Generation (2016a) About Seventh Generation. http://www.seventhgeneration.com/about-seventh-generation. Accessed 21 June 2016

Seventh Generation (2016b) Mission + action. http://www.seventhgeneration.com/corporate-consciousness-reports. Accessed 22 June 2016

Seventh Generation (2016c) 2015 corporate consciousness report. http://www.seventhgeneration.com/sites/default/files/2015_seventh_generation_corporate_responsibility_report.pdf. Accessed 20 June 2016

Shaich R (2011) Corporations must become socially conscious citizens. https://hbr.org/2011/10/corporations-must-become-socia. Accessed 10 May 2016

Siegel D (2010) Mindsight: transform your brain with the new science of kindness. Oneworld, Oxford

Simpson C (2015) Patagonia mobile worn wear tour: if it's broke, fix it. http://www.patagoniaworks.com/press/2015/3/31/patagonia-mobile-worn-wear-tour-if-its-broke-fix-it. Accessed 17 June 2016

Sisodia R, Sheth J, Wolfe D (2014) Firms of endearment: how world-class companies profit from passion and purpose. Pearson, New York

Sivachithappa K (2014) Equality and sustainable human development—issues and policy implications. Lulu, Raleigh, NC

Smith K (2014) Emile Durkheim and the collective consciousness of society: a study in criminology. Anthem, London

Southwest Airlines (2016a) Southwest corporate fact. http://www.swamedia.com/channels/Corporate-Fact-Sheet/pages/corporate-fact-sheet. Accessed 16 June 2016

Southwest Airlines (2016b) About Southwest. https://www.southwest.com/html/about-southwest/index.html?int=. Accessed 16 June 2016

Southwest Airlines (2016c) Culture. https://www.southwest.com/html/about-southwest/careers/culture.html. Accessed 16 June 2016

Southwest Airlines (2016d) 1990 to 1994. http://swamedia.com/pages/1990-to-1994?l=en-US. Accessed 16 June 2016

Statistisches Bundesamt (2013) Die wirtschaftliche Bedeutung kleiner und mittlerer Unternehmen in Deutschland. https://www.destatis.de/DE/Publikationen/WirtschaftStatistik/UnternehmenGewerbeanzeigen/BedeutungKleinerMittlererUnternehmen_12014.pdf?__blob=publicationFile. Accessed 20 June 2016

Statistisches Bundesamt (2016) 13. Koordinierte Bevölkerungsvorausberechnung für Deutschland. https://www.destatis.de/bevoelkerungspyramide/#!y=2016&v=2. Accessed 20 Jun 2016

Strauss W, Howe N (1992) Generations: the history of America's future, 1584 to 2069. William Morrow, New York

Tate C (2015) Conscious marketing: how to create an awesome business with a new approach to marketing. Wiley, Milton, QLD

The Container Store (2016) Our employee first culture. http://standfor.containerstore.com/putting-our-employees-first. Accessed 17 June 2016

Thompson C (2007) The see-through CEO. http://www.wired.com/2007/04/wired40-ceo/. Accessed 28 June 2016

Tindell T (2011) The Container Store's employee focused culture. https://www.youtube.com/watch?v=IKLuVd3zCPA. Accessed 5 May 2016

Trehan M, Trehan R (2009) Government and business. V.K., New Delhi

UK Office for National Statistics (2016) UK environmental accounts: how much material is the UK consuming?. http://www.ons.gov.uk/economy/environmentalaccounts/articles/ukenvironmentalaccountshowmuchmaterialistheukconsuming/ukenvironmentalaccountshowmuchmaterialistheukconsuming. Accessed 3 May 2016

Universum (2015) Understanding a misunderstood generation: the first large-scale study of how millennial attitudes and actions vary across the globe, and the implications for employers. http://www.headfoundation.org/userfiles/publication_reports/KC10240_Universum_Millennials.pdf. Accessed 8 June 2016

Velian (2016) Velians Weg der Holakratie mit Conscious Business. http://www.velian.de/2016/04/11/velians-weg-der-holakratie-mit-conscious-business/. Accessed 13 June 2016

Velmans M (2009) Understanding Consciousness. Routledge, London

Venkatesan R (2015) Confronting corruption. http://www.mckinsey.com/business-functions/organization/our-insights/confronting-corruption. Accessed 18 June 2016

Vieten C, Amorok T, Schlitz MM (2006) I to we: the role of consciousness transformation in compassion and altruism. Zygon J Relig Sci 41(4):915–932. https://doi.org/10.1111/j.1467-9744.2006.00788.x

Voola R, Carlson J, West A (2004) Emotional intelligence and competitive advantage: examining the relationship from a resource-based view. Strateg Change 13(2):83–93. https://doi.org/10.1002/jsc.667

Weiner J (2012) 10 leadership lessons from Jeff Weiner, CEO of LinkedIn. http://www.forbes.com/sites/jeannemeister/2012/09/10/10-leadership-lessons-from-jeff-weiner-ceo-of-linkedin/#6e869b8452b1. Accessed 12 June 2016

Weiner J (2016) LinkedIn CEO: here's why I sold the company to Microsoft. http://time.com/4366106/linkedin-microsoft-why/. Accessed 15 June 2016

Wellner K (2001) Shareholder-Value und seine Weiterentwicklung zum market adapted Shareholder-Value approach: Entwicklungslinien, Probleme und Lösungsansätze einer Shareholder-value- orientierten Unternehmensführung. Tectum, Marburg

Wheeler BO (1962) Business. An introductory analysis. Harper, New York

Whole Foods (2016a) Our core values: what's truly important to us as an organization. http://www.wholefoodsmarket.com/mission-values/core-values. Accessed 14 June 2016

Whole Foods (2016b) Company info: America's healthiest grocery store. http://www.wholefoodsmarket.com/company-info. Accessed 14 June 2016

Whole Foods (2016c) Declaration of interdependence. http://www.wholefoodsmarket.com/mission-values/core-values/declaration-interdependence. Accessed 14 June 2016

Whole Planet Foundation (2016) About the foundation. https://www.wholeplanetfoundation.org/about/. Accessed 21 June 2016

Wigglesworth C (2004) Spiritual intelligence assessment. http://www.deepchange.com/BB1130402SampleReport.pdf. Accessed 23 June 2016

Wilber K (2006) Foreword. In: Kofman F (ed) Conscious business: how to build value through values. Sounds True, Boulder, CO

World Bank (2016a) Health expenditure. http://data.worldbank.org/indicator/SH.XPD.TOTL.ZS?end=2014&start=2014&view=map. Accessed 21 June 2016

World Bank (2016b) Internet users. http://data.worldbank.org/indicator/IT.NET.USER.P2?end=2014&start=1990&view=chart. Accessed 4 July 2016

World Business Council for Sustainable Development (1998) Meeting changing expectations: corporate social responsibility. http://www.wbcsd.org/pages/edocument/edocumentdetails.aspx?id=82&nosearchcontextkey=true. Accessed 11 June 2016

World Commission on Environment and Development (1987) Our common future. www.un-documents.net/our-common-future.pdf. Accessed 8 Mar 2016

World Economic Forum (2011) Faith and the global agenda: values for the post-crisis economy. http://www3.weforum.org/docs/WEF_GAC_FaithValuesReport_2011.pdf. Accessed 4 Jul 2016

Zappos (2016a) About Zappos: company history. http://www.zappos.com/c/about-zappos. Accessed 12 June 2016

Zappos (2016b) About Zappos culture: our culture. http://www.zappos.com/core-values. Accessed 12 June 2016

Zappos (2016c) Higher purpose. https://www.zapposinsights.com/about/zappos/higher-purpose. Accessed 12 June 2016

Zimbalist A, Sherman H (1984) Comparing economic systems: a political-economic approach. Academic, Orlando, FL

Zohar D, Marshall I (2004) Spiritual capital: wealth we can live by. Berrett-Koehler, San Francisco, CA

Chapter 3
Research Framework for the German Case Studies

3.1 Study Design

This study represents a cross-sectional study, which is based on a non-experimental study design. The data was gathered by analyzing company documents, conducting guided interviews, and including existing findings from related studies and articles.

3.2 Measuring Instruments

A qualitative content analysis was conducted to gather information regarding Conscious Business principles which are externally communicated by selected companies. The content sources are mainly represented by various company reports, articles, company websites, and further sources like press releases. This analysis was conducted to collect information regarding the elements higher purpose, conscious leadership, conscious culture, stakeholder integration, and further aspects which are related to the concept of conscious business. Afterwards, the findings were composed in an individual case study for each company. The qualitative content analysis was conducted according to the principles of Mayring (2010) which are emphasized in Fig. 3.1.

The selection criteria and level of abstraction have been defined in such a way, that a direct link to the elements higher purpose, conscious leadership, conscious culture, stakeholder integration and other conscious business aspects could be identified. The gradual development of categories was done for all mentioned factors and resulted in an individual category for each item. This procedure counteracts the bias to prematurely sort the content to a summarizing category. The formative review of reliability was conducted after analyzing 30% of the company's reports and documents. Similar categories had been compared and bundled to a new category to avoid redundancies. This step was repeated in the

© Springer International Publishing AG 2018

N.J. Stahlhofer et al., *Conscious Business in Germany*, CSR, Sustainability, Ethics & Governance, https://doi.org/10.1007/978-3-319-69739-0_3

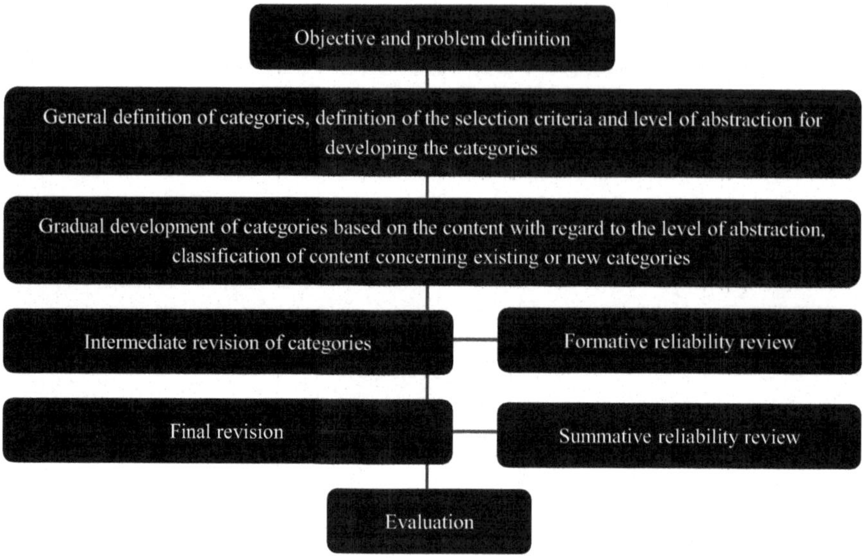

Fig. 3.1 Qualitative content analysis process (author's diagram)

final revision to evaluate the existing categories with a summative reliability review.

If an item could not be classified to one of the bundled categories, it was moved to the category "other". As a final step, a quantitative analysis was conducted to gain insights regarding the relevance of the respective category. Guided interviews embody the second measuring instrument which was used in this study. The development of the interview guideline was oriented on previous findings from the reviewed literature. Overall, 17 questions were developed which had been sorted into five different categories. The first category contained information about the areas of responsibility of the interview partner. The following questions were linked to the definition and attitude regarding the concept of conscious business. The third part was concerned with experiences that are linked to the internal and external implementation of principles in the company. The following category addressed explicit practices of how the dimensions, *higher purpose*, *conscious leadership*, *conscious culture*, and *stakeholder orientation* are carried out in the respective company. The final category contained three questions regarding the possible future development of the concept. Figure 3.2 summarizes the chronological process regarding the conducted interviews.

The interviews were analyzed according to Lamnek (2010). Figure 3.3 outlines the main parts of the process.

The transcripts were reviewed regarding its content and spelling as well as logical connections between the different paragraphs. The corrected version was sent to the interview partners to review the content. After receiving the corrected versions, the interview transcripts were analyzed individually. Core aspects and

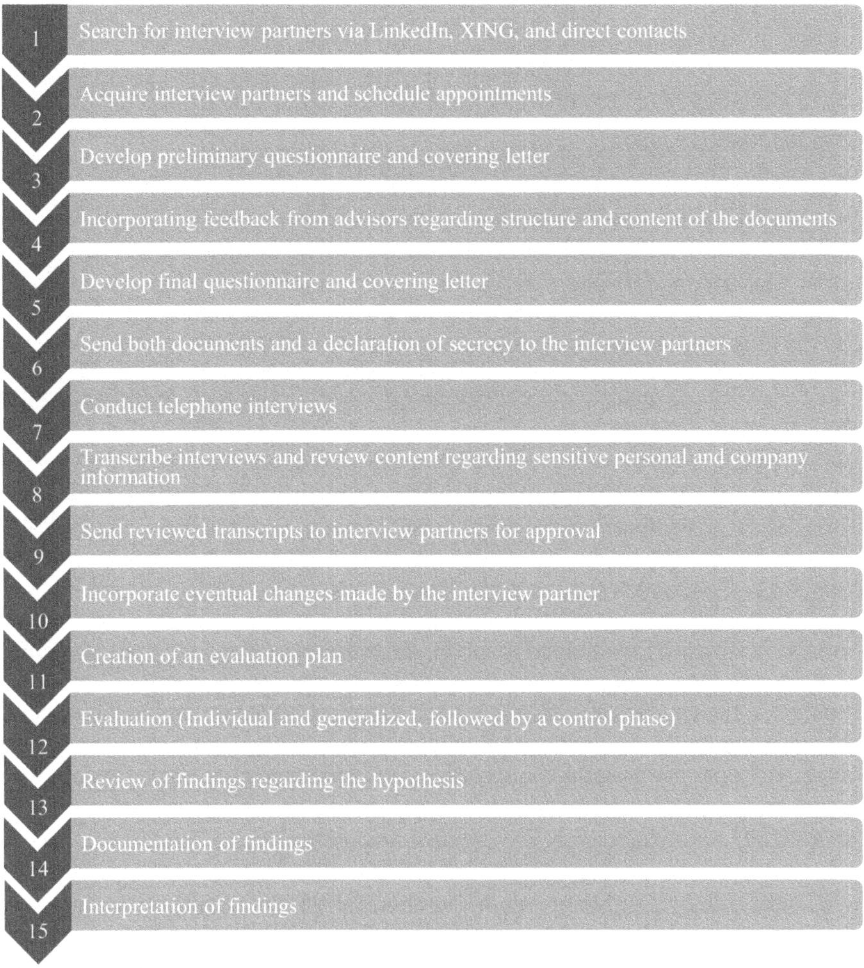

Fig. 3.2 Interview process (author's diagram)

Fig. 3.3 Interview evaluation process (author's diagram)

content fragments were paraphrased to bundle individual statements. In the following step, the paraphrased statements were analyzed according to the qualitative content analysis as described in Fig. 2.20.

The analyzed information was classified and coded according to the questions numbers in the questionnaire. A generalized coding system could not be applied to ensure the individuality of statements (Steinke 2008). Based on the results of the individual analysis, a statement matrix was created in the step of the generalized analysis. This matrix compiles information regarding the interview person, line number of the individual transcript, the paraphrased statement, as well as the respective category. During the control phase, the initial transcripts were compared to the generalized matrix. This procedure is necessary to minimize evaluation errors due to subjective content assessment and misinterpretations.

3.3 Quality Criteria

With regard to the analysis of qualitative data, Barbour (2014) notes that there is no generalized set of quality criteria as it exists for the interpretation of quantitative data. In order to ensure a consistent set of data, Schreier (2014) emphasizes the importance of an incremental development of coding schemes and processes in analyzing qualitative data. With reference to this book, the evaluated content was analyzed by using the process according to Fig. 3.2. The categories were built in a *concept-driven* approach which means that the categories are based on previous knowledge from a theoretical foundation (Schreier 2014). In addition, the selected categories were first assessed in a *trial coding phase* to compare the findings without discarding sub-categories. In the following step, the codes were modified according to their similarities and variances. The codes were created according to findings in the theoretical framework to cover relevant themes and link connected aspects according to the process of Rubin and Rubin (2005). The quality criterion of objectivity does not have the same connotation with qualitative data as it does with quantitative data (Lamnek 2010). With regard to qualitative data, objectivity can be understood as an intersubjective replicability of findings. This aspect can be achieved by documenting the process of data collection and analysis which is described in Sect. 3.2. Moreover, the combination of guided interviews and qualitative content analysis represent valid instruments to generate and evaluate qualitative data (Bortz and Döring 2006). Hence, this procedure allows the discovery of variations and meanings and helps to reflect the complexity of qualitative data (Rubin and Rubin 2005).

3.4 Sample Design

The companies for the following case studies were selected in an incremental process. At first, all German companies which are listed as a B Corporation or a member of the Economy for the Common Good were gathered for further revision. The analysis was conducted by searching for similarities with regard to the elements higher purpose, conscious leadership, conscious culture and management, and stakeholder integration. Among these elements, the aspect of higher purpose represented the core selection criterion as it represents the core of every conscious business (Mackey and Sisodia 2014). The search for correlating content was conducted by analyzing company information, books, articles, and studies. In addition to companies, which are listed as B Corporations or members of the Economy for the Common Good, further businesses were evaluated regarding their human-centric approach.

Moreover, studies and rankings with reference to sustainability and ethical company behavior were considered to ensure different approaches in finding potential companies for the following case study. In order to depict a variety of perspectives, four companies were selected which differ in size and industry. These companies are dm drogerie markt, VAUDE, Polarstern, and the GLS Bank. Afterwards, the founders or managing directors of each company were contacted by e-Mail to schedule a personal or telephone interview. In addition, an executive from the automotive industry who is referred to as *Interview partner 1* and the founder of the Conscious Business Institute, Peter Matthies were interviewed to enable a wider range of perspectives on the implementation of conscious business principles. The interviews took place from May 27th to July 5th 2016. Five interviews were conducted via telephone and the co-founder of Polarstern, Florian Henle was interviewed in person.

3.5 Study Conduct

The case study is designed to explore the current situation of conscious business principles in selected German companies. As described in Sect. 2.6, different American companies such as Zappos, Whole Foods Market, Southwest Airlines, and LinkedIn have implemented various characteristics of the conscious business tenets. Hence, the following case study aims at identifying aspects of conscious business principles according to the research framework and the characteristics which are outlined in Sect. 2.2. The case study is embedded in the context of explanatory and exploratory research (Saunders et al. 2012). This case study combines information which derives from interviews, company documents, related articles, and findings from the literature. Hence, the mix of methods is designed to fulfill a triangulation of methods which allow a holistic view on the research phenomenon. With regard to the type of information, this case study combines qualitative and quantitative data to create a holistic image. The structure of the cases is designed in a semi-structures way. Hence, all companies are reviewed regarding the four elements of the research framework to enable a comparative understanding. In addition, every case will highlight different orientations of the company to portray the individual orientation of every company.

References

Barbour RS (2014) Quality of data analysis. In: Flick U (ed) The Sage handbook of qualitative data analysis. Sage, London, pp 496–509

Bortz J, Döring N (2006) Forschungsmethoden und Evaluation für Human- und Sozialwissenschaftler. Springer, Heidelberg

Lamnek S (2010) Qualitative Sozialforschung. Beltz, Weinheim

Mackey J, Sisodia R (2014) Conscious capitalism: liberating the heroic spirit of business. Harvard Business School Publishing, Massachusetts

Mayring P (2010) Qualitative Inhaltsanalyse. In: Flick U, von Kardorff E, Steinke I (eds) Qualitative Forschung. Rowohlt, Reinbek, pp 468–475

Rubin HJ, Rubin IS (2005) Qualitative interviewing: the heart of hearing data. Sage, Thousand Oaks, CA

Saunders M, Lewis P, Thornhill A (2012) Research methods for business students. Pearson, Harlow

Schreier M (2014) Qualitative content analysis. In: Flick U (ed) The Sage handbook of qualitative data analysis. Sage, London, pp 170–183

Steinke I (2008) Gütekriterien qualitativer Forschung. In: Flick U, von Kardorff E, Steinke I (eds) Qualitative Forschung. Ein Handbuch. Rowohlt, Reinbek, pp 319–331

Chapter 4
Case Studies: Conscious Business in Germany

As described above, this chapter analyzes four selected companies and their implementation of conscious business principles. Each company is outlined with a company profile and a brief description of how the company has developed. The conclusion at the end of each case compares the findings with the principles of conscious businesses according to the theoretical framework.

4.1 GLS Bank

The GLS Bank is the subject of the first case study (Fig. 4.1). The information regarding conscious business principles is derived from the available company information, books, and an interview with the CEO's longstanding advisor, Janina Zajic. This case emphasizes implementation of conscious business principles with a focus on value-based banking and the company's focus on purpose.

4.1.1 Company Development

The GLS Bank is a sustainable, ethical bank with a distinct focus on a conscious approach to its operations and investments. This setting has been translated into a variety of sustainability funds, micro-finance models, transparent investment guidelines and a profitable business model which has been developed over the past decades. The GLS Bank was started in Bochum by Gisela Reuther, Wilhelm Ernst Barkhoff, Albert Fink, and Rolf Kerler in 1974. The bank was founded on an anthroposophical understanding. The first branch was established in Stuttgart in 1982. Seven years later, the GLS Bank set up its first wind power fund in order to foster investments in this sector and support the expansion of renewable energy. In the same year, the GLS Bank co-founded the International Association for Investors

© Springer International Publishing AG 2018

N.J. Stahlhofer et al., *Conscious Business in Germany*, CSR, Sustainability, Ethics & Governance, https://doi.org/10.1007/978-3-319-69739-0_4

Company Profile

GLS Bank
das macht Sinn

(GLS Bank 2017a)

Founded	1974 in Bochum (Germany)
Industry	Banking
Headquarter	Bochum
CEO	Thomas Jorberg
# Employees 2016	524
Revenue 2016	€4.596 billion

Fig. 4.1 GLS Bank Company profile (authors' diagram)

in Social Economy, which is a network of financial institutions that expedite positive social and environmental change (International Association of Investors in the Social Economy 2016). In 2003, the GLS Bank took over the Ökobank. Two years later, it founded a micro-finance institute and helped to establish the Institute for Social Banking.

In June 2008, the GLS Bank integrated the Integrabank in Munich and in July it opened another branch in Berlin. In 2008, the GLS Bank held over one billion € in assets under management. One year later, the GLS Bank formed the Global Alliance for Banking on values together with other banks worldwide to strengthen and expand the role of alternative banks. In the same year, the spokesman of the management board, Thomas Jorberg, received an award for his responsible company leadership and the consistent sustainability management of the GLS Bank. The GLS Bank celebrated its 40th anniversary in June 2014. One year later, the bank passed the four billion € threshold in assets under management and was nominated as Bank of the Year in Germany for the sixth time in a row. In addition to a variety of national awards in the areas of sustainability, responsible investments, fairness and leadership, the GLS Bank was also recognized as Europe's most sustainable bank in 2013. Its customer services include current accounts, retirement plans, asset management, investment services and various financing models (GLS Bank 2016a). The GLS Bank supports investment decisions by supporting assessment of the object's security, disposability and return as well as its impacts on a social, ecological and the real economy (Jorberg 2016a).

4.1.2 Value-Based Banking

The GLS Bank is convinced that money should benefit sustainable development for mankind and the environment. The bank sees itself as a pioneer in social-ecological banking and emphasizes that the money invested is used in a purposeful and transparent way. Sustainability is a uniting value for connecting money and people (GLS Bank 2014a, 2016b). The bank focuses on aligning the interests and values of its clients with their investment capabilities and opportunities on a trustful basis.

This has been the bank's understanding ever since it was established because its founders were strongly focused on implementing an anthroposophical approach to banking which is based on humanity. This mindset aimed at creating a banking system which makes it possible for people to invest their savings consciously for the sake of creating a benefit for society (Kerler 2011a). Kerler (2011b) defines the bank as an enabler of altruistic behavior for individuals and a multiplier for promoting these efforts. Hence, money is not perceived as an end in itself, but as a medium for realizing social and ecological goals (Kampffmeyer and Knopf 2011).

The GLS Bank conducted a brand excellence program to clarify the values that align the bank with its clients. This program was necessary because the bank had been growing significantly in terms of its members, customers, and employees. It identified seven values that unite the interests of the bank with the interests of its clients (Zajic 2016). These values are shown in Fig. 4.2.

Conscious is a core value for the GLS Bank because it is the principle which guides the way in which the bank operates. The value *human* should be manifested and visible in every internal and external interaction with the bank and its stakeholders. *Collaborative* relates to integration of stakeholder interests and the international cooperation networks which the GLS Bank has established such as, for example, the Global Alliance for Banking on Values. *Future-oriented* expresses the long-term orientation of the bank and challenges every member of the organization to constantly innovate and explore opportunities. Hence this value is directly oriented to economic, societal, and environmental changes which can be supported by the GLS Bank. For example, the bank recently started an initiative which promotes the quality of journalistic work. Although this topic is not directly related to the bank's business activities, it certainly is a field to which the bank wants to contribute in favor of society. *Purposeful* represents the bank's aim to align every activity to a purpose which benefits its stakeholders. This value is also linked to *transparent* communication with clients in terms of how their investments are used. The final value is *reference*. This characteristic relates to the bank's self-concept of

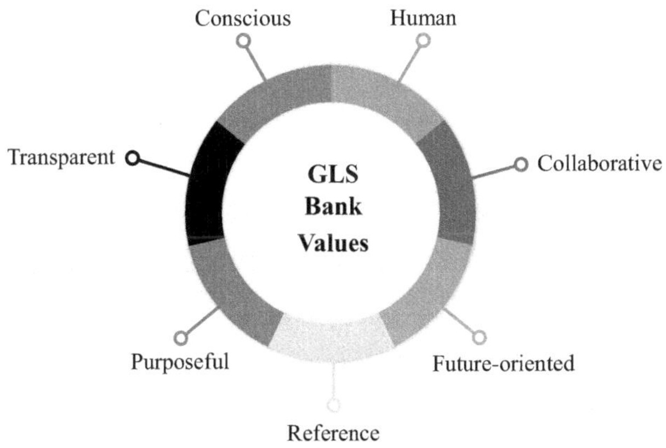

Fig. 4.2 Seven values at the GLS Bank (authors' diagram)

being the first bank which comes to clients' minds in regard to ethical and sustainable banking. The GLS Bank aims to establish these values with its clients, members and partners in Germany and to promote the idea of ethical and sustainable banking worldwide (Zajic 2016). For this reason, the bank has established various collaborations with other banks that pursue a similar approach (Kerler 2011a). One of these networks is the Global Alliance for Banking on Values (GABV), which is a coalition of banks and leaders around the world who are committed to advancing positive changes in the banking industry. The collective goal is to achieve a systemic change that leads to a higher degree of transparency, sustainability and positive impact of investments (Global Alliance for Banking on Values 2016). Figure 4.3 highlights the global members of this network and their geographic distribution.

Weber and Scheck (2012) emphasize that the GLS Bank is a pioneer and leader in the segment of value-based banking. This aspect can be identified by an increasing number of people who trust their business model in Germany. The pioneer role in social investment and ethical business operations is also visible in the bank's engagement in the area of social entrepreneurship which is a growing trend (Bertelsmann Stiftung 2016). With reference to this development, UBS (2015) recognizes an ongoing change of values in society to a more sustainable investing culture. As a result, responsible investing has both grown in terms of assets under management as well as in popularity among customers and shows an increasing demand for ethical and conscious investing (Aburdene 2005). According to KPMG (2015), the European Responsible Investment Fund Assets under Management increased from 200 billion € in 2010 to 372 billion € in 2014. This steady increase in sustainable investing was led primarily by France with a Compounded Annual Growth Rate (CAGR) of 213% from 2013 to 2015 (43 billion € in 2015), followed by the Netherlands with a CAGR of 36% (37 billion € in 2015) and Switzerland, CAGR of 29% (21 billion in 2015). Germany is ranked in fifth place with a CAGR of 41% and a total volume of eight billion € in 2015 (Eurosif 2016).

The following paragraphs outline the manifestation of the research model at the GLS Bank.

4.1.3 Higher Purpose

The GLS Bank defines the company's higher purpose as the *aim to distribute money where it is needed*. This understanding has been an inherent core of the company since it was founded. Although the bank has developed in terms of its business structure activities, its purpose has remained the same. With regard to the employees, it is important that candidates match with the bank's philosophy.

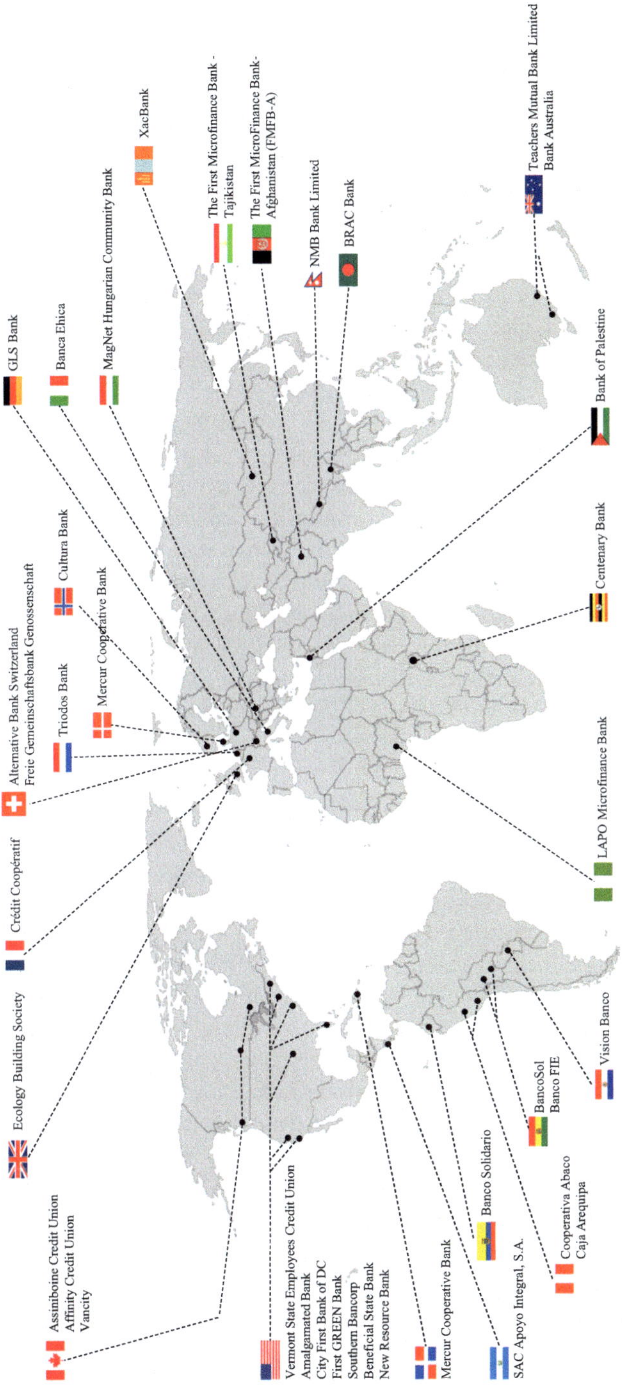

Fig. 4.3 Members of the Global Alliance for Banking on Values (authors' diagram)

4.1.4 Stakeholder Integration

Integrating stakeholders on the premise of fulfilling the company's purpose is a core characteristic of the GLS Bank according to Janina Zajic. The GLS Bank considers stakeholders as organizations or persons that are affected by its activities, offers and services. The bank's members, employees, clients, partners, as well as those in the communities in which they operate are the bank's most important stakeholders. The GLS Bank has been awarded the *German Award for Excellence* in 2015 for its responsible treatment of stakeholders in the category *fair business* (GLS Bank 2015a). With reference to its social and environmental orientation, the GLS bank highlights the importance of creating a collaborative environment based on trust and transparent interactions (GLS Bank 2013). The GLS Bank perceives the stakeholder management approach as superior to the shareholder-value approach. Investors are perceived as a part of a larger network that is based on collaboration rather than unbalanced preference. Promoting dialogue among stakeholders is achieved by various formats such as the *general assembly*, which is held once a year. At this event, stakeholders are encouraged to contribute to the development of the bank and discuss their ideas regarding current and future topics (GLS Bank 2015b). Employees can join several formats at work, such as *Montagmittag* [Monday Noon] or *Markt der Möglichkeiten* [Market of Possibilities] where discussions are held to shape the direction of the GLS Bank. External stakeholders are invited to these events to present their initiatives, to work on projects together, and to generate new ideas. In addition, the GLS Bank supports initiatives and projects in terms of financial and logistic needs.

These measures aim at building a better network and generating a greater good for society (Zajic 2016). Overall, Götz Werner, founder of the dm drogerie markt and member of the supervisory board of the GLS Bank, notes that the GLS Bank gives its members the opportunity to shape the activities of the bank inasmuch as the bank's structure allows for a high degree of participation. In addition, the members are able to decide on specific areas which should be supported in the future by voting for them in general assemblages (Dohmen 2011). Transparency is a key element of the GLS Bank's stakeholder integration, which is visible in the bank's efforts to publish all investments. By doing so, the GLS Bank aims at building trust among its customers and members and visualizing its impacts on ecologic and social dimensions. Therefore, every four months it releases a document called *Bankenspiegel* which summarizes its current investments with a declaration of the specific business or institution and the amount invested (GLS Bank 2016c). This customer journal was first released in 1980 and has served as a key activity in establishing transparency regarding the bank's operations.

4.1.5 Conscious Leadership, Culture, and Management

Janina Zajic states that besides the brand excellence initiative, the GLS Bank has also conducted a program on leadership excellence. This program was needed to create a unified leadership perspective because the bank had been growing steadily and many new leaders had joined the company. Integrating the employees was an integral element in this process of working collaboratively on a joint mindset of leadership and values. In addition, the bank established a so called *Führungswerkstatt* [leadership workshop] in 2014 to focus on developing leaders with regard to future developments of the bank and a common understanding of leadership. The creation of interdisciplinary teams is another element which has been designed to encourage employee flexibility and support the creation of internal initiatives. According to Zajic (2016), enabling employees to work with a high degree of self-efficacy with the support of leaders is an important goal for the GLS Bank. Another element that has been established is collegial coaching, which also relates to the foregoing initiatives. Mindfulness plays an important role at the GLS Bank and in its understanding of leadership. Consequently, perceiving each leader on four different dimensions is an essential component. Each leader is seen as an *individual, in his or her role as a leader, in his or her role in the organization*, and *in terms of his or her impact on the environment and society*. In collaboration with other companies, the GLS Bank also conducts so called *Lernreisen* [study trips] to benefit from the perspectives and approaches of others. With reference to growth during the past 40 years, Janina Zajic notes that these measures are likely to contribute to a clear structure internally. Young leaders, in particular, appreciate the defined direction of leadership. Although there are still areas in which this transformation could be improved, its direction will be reinforced in the future. The corporate culture is based on honesty, trust, respect, inclusion and a positive holistic image of humanity which expresses the underlining basis. While the GLS Bank does not have a works council, it does have a so called *Vertrauenskreis* [trust committee] which is elected and defined in accordance with its employees. The GLS Bank focuses on the quality of work rather than on quantitative key performance indicators. Therefore, performance reviews are not based on sales objectives. The GLS Bank values the way customers and members are supported in finding investment opportunities without the adviser being under pressure to sell inappropriate products and services (Zajic 2016).

4.1.6 Company Performance

The GLS Bank's economic activity focuses on purpose rather than on profits. The bank aims to continue developing the company in a stable and sustainable way by supporting social and ecological projects that benefit society as a whole. The GLS Bank has been growing continually since it was founded and has announced its intention to expand its business activities even more in a social and ecological direction. Figure 4.4 clarifies the development of total assets and customer deposits at the GLS Bank from 2012 to 2015 (GLS Bank 2014b, 2016d).

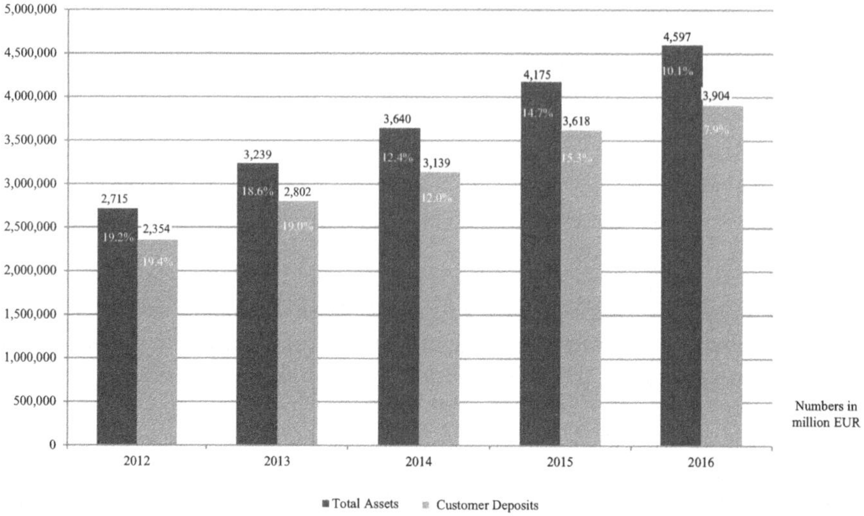

Fig. 4.4 Development of total assets and customer deposits at the GLS Bank (authors' diagram)

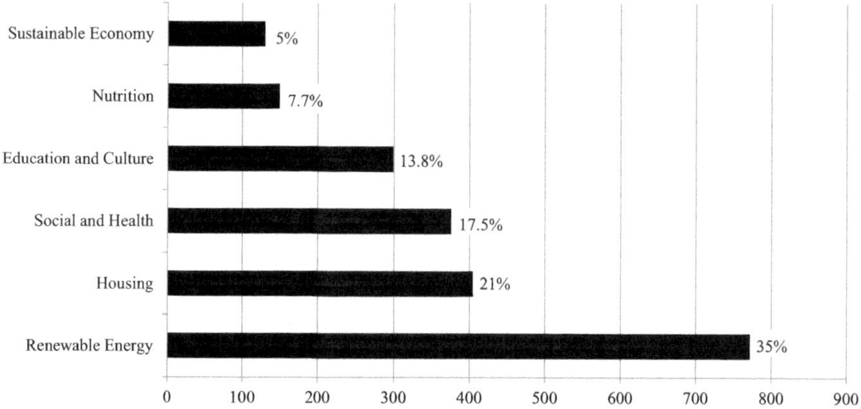

Fig. 4.5 Distribution of investments by sector at the GLS Bank in 2015 (authors' diagram)

With reference to its total assets, the GLS Bank was able to reach a new high of 4.597 billion € in 2016. This development translates into a Compound Annual Growth Rate of 11.1% and overall growth of 69.29% since 2012. In 2016, the GLS Bank had around 211,000 customers and 46,000 members which used the services of the bank and invested their money in various projects with a social or ecological orientation. The constant growth in members has continued in 2016 and resulted in a total number of 46,313, which translates into an increase of 10.3% (GLS Bank 2016e, 2017b). A total of 2452 million € was invested in 24,600 businesses by the end of 2016 (GLS Bank 2016f). In addition, the GLS Bank's loan default rate accumulated at the very low rate of 0.1% in 2016 (Jorberg 2016b). Figure 4.5 outlines the distribution of all investments made by the GLS Bank in 2015 over various sectors in percent.

Table 4.1 Customer satisfaction among German banks (authors' table)

Rank	Bank	Overall score	Quality rating
1	GLS Bank	88.7	Very Good
2	Ethik-Bank	88.3	Very Good
3	Comdirect Bank	86.2	Very Good
4	Bank für Kirche und Diakonie—KD-Bank	85.2	Very Good
5	1822direkt	84.7	Very Good
6	PSD Bank Nürnberg	84.1	Very Good
7	ING-Diba	83.9	Very Good
8	Sparda Bank Südwest	83.2	Very Good
9	DKB Deutsche Kreditbank	83.1	Very Good
10	Wüstenrot direct	82.1	Very Good
11	PSD Bank Köln	82.0	Very Good
12	PSD Bank Westfalen-Lippe	81.7	Very Good
13	Kreissparkasse Köln	80.9	Very Good
14	Sparda-Bank München	80.4	Very Good
15	Consorsbank	80.3	Very Good
16	BB Bank	78.5	Good
17	Norisbank	78.3	Good
18	Sparda-Bank Berlin	71.6	Good
19	Targobank	68.9	Satisfactory
20	Postbank	68.4	Satisfactory
21	Commerzbank	67.1	Satisfactory
22	Hypovereinsbank	63.6	Satisfactory
23	Deutsche Bank	62.4	Satisfactory

According to the anonymous job rating portal kununu (2016), the GLS Bank received an average rating of 4.67 out of 5. Since the average job portal rating in the banking industry is 3.35, this is an exceptionally good result. With reference to customer satisfaction, the GLS bank was nominated as Bank of the Year 2015, based on a survey of 16,000 participants (Deutsches Institut für Service-Qualität 2015, Table 4.1).

4.1.7 Conclusion

The GLS Bank combines purpose with profits on the base of human flourishing and active support for social and ecological well-being worldwide (Stein 2010). The company is founded on an anthroposophical philosophy which translates into its human-centric, sustainable approach regarding its business activities. Kofman (2006) sees a human-centric approach as an inherent aspect of conscious businesses which is present at the GLS Bank. Their guiding principle states that they focus on the nature of life, peaceful coexistence of cultures, individual freedom and personal responsibility. In addition, the bank emphasizes its focus on a holistic view of humans which encompasses body, mind, and soul. This orientation directs their behavior on the premise of contributing to a world that is livable for present and future generations. Their understanding is closely linked to a generational world-view which has been identified as a core characteristic of conscious businesses by Lerner et al. (2003) and Sisodia et al. (2014). The GLS Bank's higher purpose is reflected in its aim to distribute money where it is needed. This purpose is a guiding direction for the bank to promote positive change for society and the environment globally. With reference to the classification by Mourkogiannis et al. (2008), Mackey and Sisodia (2014), the GLS Bank relates to the ideal of *the good*. The GLS Bank is driven by its values *conscious*, *human*, *purposeful*, *future-oriented*, *collaborative*, *reference*, and *transparent*. These values are cultivated both internally and externally, and should be visible in every interaction (Zajic 2016). Moreover, the bank focuses on acting with a high degree of integrity which directs its investment decisions. According to its financing principles, the GLS Bank only supports businesses and initiatives that are in accordance with social and ecological goals. Hence, all investments that could be connected to human rights violations, non-renewable energies, the defense industry and other harmful categories are not supported (GLS Bank 2016g). To establish this understanding worldwide, the GLS Bank has established various networks for exchanging knowledge and creating mutual benefit through international partnerships with other banks. One of these networks is the *Global Alliance for Banking on Values* which is designed to advance positive changes in the banking industry (Global Alliance for Banking on Values 2016). Creating a positive systemic impact on stakeholders based on collaboration is a guiding principle of the GLS Bank regarding the integration of various interests. Promoting such a mindset regarding stakeholder integration is named as a distinct characteristic of conscious businesses according to Sisodia et al. (2014) and Friedman and Miles (2006). The culture and leadership style at the GLS Bank has evolved over the past 40 years and is focused on enabling self-efficacy, flexibility, and personal development of employees under collegial supervision of leaders. In addition, the bank focuses on different dimensions of an individual in the context of his or her individuality, personal role, and role within the organization and in society. Mackey and Sisodia (2014), Secretan (2001), and Kofman (2006) view these characteristics as essential for leadership in conscious organizations. In terms of its culture and management approach, the GLS Bank applies the principles of TACTILE which are postulated by Mackey and Sisodia (2014) as essential for conscious cultures. In terms of the company's performance, the GLS Bank has been steadily growing since its foundation and had around 211,000 customers and 46,000 members by the end of 2016. The total assets and customer deposits have increasingly grown by double digit percentages during the past 4 years.

With reference to the bank's long-term stability, Thomas Jorberg emphasizes that the GLS Bank did not lose any money during the financial crisis in 2008 because it deliberately avoids speculative and abstract investments (Jensen 2011). The GLS Bank has been nominated as Bank of the Year in Germany six times in a row since 2010 and achieved the highest possible rating in customer satisfaction

(GLS Bank 2016h). Furthermore, employee satisfaction lies far above the industry average according to the portal kununu (2016). The GLS Bank displays various indicators that relate to superior performance and have been named a key characteristic of conscious businesses according to Mackey and Sisodia (2014). Overall, the GLS Bank incorporates multiple factors of the research framework and thus can be seen as a representative of conscious business principles in Germany.

4.2 VAUDE

VAUDE is the subject of the second case study (Fig. 4.6). The information regarding conscious business principles is derived from available company information, books, and an interview with the CEO, Antje von Dewitz. This case underscores the importance of implementing conscious business principles with a focus on the company's sustainability efforts and stakeholder management.

4.2.1 Company Development

Albrecht von Dewitz founded VAUDE in 1974, when he started the business in a barn by selling outdoor equipment such as backpacks and sports fashion. It is still a 100% family-owned business. The production site in Obereisenbach, a village near Lake Constance in the south-west of Germany, was opened in 1980 and allowed the company to produce on a larger scale. More product categories such as bike equipment and other lifestyle items were added to the portfolio over time. This solidified the company's development. Today, more than half of the village's area is covered by VAUDE's production sites and offices, so that the locals call their village "VAUDE-City". As part of its social ambitions, VAUDE opened an on-site child care center in 2001 to offer additional support for employees with families and at the same time to improve the regional infrastructure. In the same year, the company was certified with the ecological *bluesign* standard as the first sporting goods

Company Profile

VAUDE

(VAUDE 2017a)

Founded	1974 in Tettnang (Germany)
Industry	Consumer Products
Headquarter	Tettnang
CEO	Antje von Dewitz
# Employees 2016	490

Fig. 4.6 VAUDE Company profile (authors' diagram)

manufacturer. In 2006, VAUDE bought the climbing gear brand Edelrid. In 2009, the company founder's daughter, Antje von Dewitz, took over as CEO. As part of its sustainability focus, VAUDE launched an individual sustainability program to guarantee a high standard of ecological products and manufacturing processes.

Further certifications and sustainability partnerships were established in the years thereafter to support the company's vision of becoming Europe's most sustainable outdoor goods manufacturer. These measures were also supported by the "climate-neutral" production site in Germany implemented in 2012. Three years later, VAUDE joined the Economy for the Common Good movement as the first company in the outdoor industry to clarify the link between business and society (VAUDE 2016a). VAUDE's customers are to a large extent environmentally conscious consumers who understand and support its philosophy and mission. In fact, some 10% of VAUDE's customers are "A-customers", meaning that they account for roughly 70% of total sales volume. In 2015 mountain sports equipment contributed 58% to overall revenue while the share of bike sports equipment was 35% and of "packs 'n bags" 7%. VAUDE has diversified beyond mountain sports in order to become less dependent on just one line of business. VAUDE is especially well known in Germany for its very first product, the backpack (which contributes 26% to overall revenue) as well as the disproportionally growing segments of outdoor-related clothes (62% of revenue). Smaller revenue shares come from tents (4%), shoes (2%) as well as sleeping bags (2%). Most of its revenues are generated in Germany (73.1%) while neighboring European countries account for a share of up to 4.7% (Switzerland). VAUDE is available in most relevant retail stores in Germany, where 70% of all revenue is generated. E-commerce became an important cornerstone of VAUDE's sales in recent years. Online retail stores and the vaude.com online shop contributed 22% of total sales in 2015. VAUDE has combined its paths to sustainability and e-commerce and provides a "VAUDE Second Use Shop" together with eBay on its platform. As described below, VAUDE tries to make its supply chain as transparent as possible. The company works closely with its suppliers, especially in terms of education, quality and sustainable development. Most producers are in China (46%) and Vietnam (26%). The suppliers of materials are mainly located in Taiwan (47%), mainland China (16%), Italy and South Korea (11% each). At the same time, VAUDE tries to put a stronger focus on production "made in Germany". After the original workshop burned down in 2015 because it was struck by lightning, today around 100,000 backpacks and bike bags are produced in the recently rebuilt workshop at Tettnang-Obereisenbach. This means, for example, that 20% of VAUDE's backpacks are now made in Germany.

4.2.2 *Focus on Sustainability*

This family-owned business has been constantly developed towards its vision of becoming Europe's most sustainable outdoor goods manufacturer (VAUDE

2016a). The company's vision is based on an inherent consciousness regarding environmental impacts of business and the necessity of integrating economical, ecological, and social needs in a holistic concept (VAUDE 2014). Wuczkowski (2015) notes that VAUDE aims at being a pioneer and wants to inspire other companies to join this movement. VAUDE perceives sustainability as a complex network of different dimensions which are described as a *green thread* for the company. The sustainability goals encompass economic self-sufficiency, transparent and authentic behavior, high environmental and social standards in the value chain, and satisfied employees (VAUDE 2016c). The VAUDE ecosystem of production activities is outlined in Fig. 4.7.

The company uses production sites in Germany, China, and Vietnam, and cooperates with other product suppliers in Asia and Europe. With regard to production standards and fair treatment, VAUDE applies Western production standards at their sites abroad as a part of their fair treatment policy (Fitzner 2009). Antje von Dewitz (2016) remarks that VAUDE constantly applies strict standards with regard to their products and production methods, and aims at establishing a higher standard in the industry worldwide. The VAUDE *Green Shape* classification represents a company-specific sustainability standard which was first introduced in 2010. It is a core part of the VAUDE eco-collection and measures the environmental impact linked to a product. This program was established because there was no holistic classification system for products in this segment. Referring to this, VAUDE emphasizes the negative impact on the environment linked to the use of polyfluoride and perfluoride chemicals because these components cannot be recycled properly. VAUDE aims to produce its entire collection without any of these components by 2020. By the end of 2015, the company had already achieved

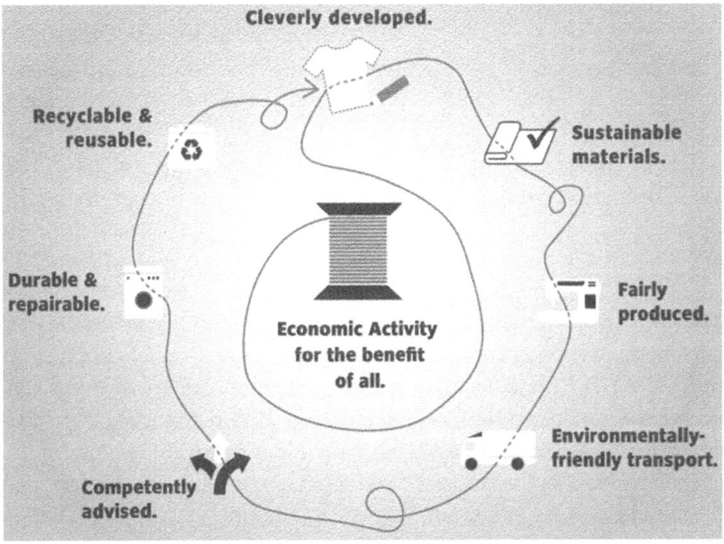

Fig. 4.7 VAUDE ecosystem (VAUDE 2016c)

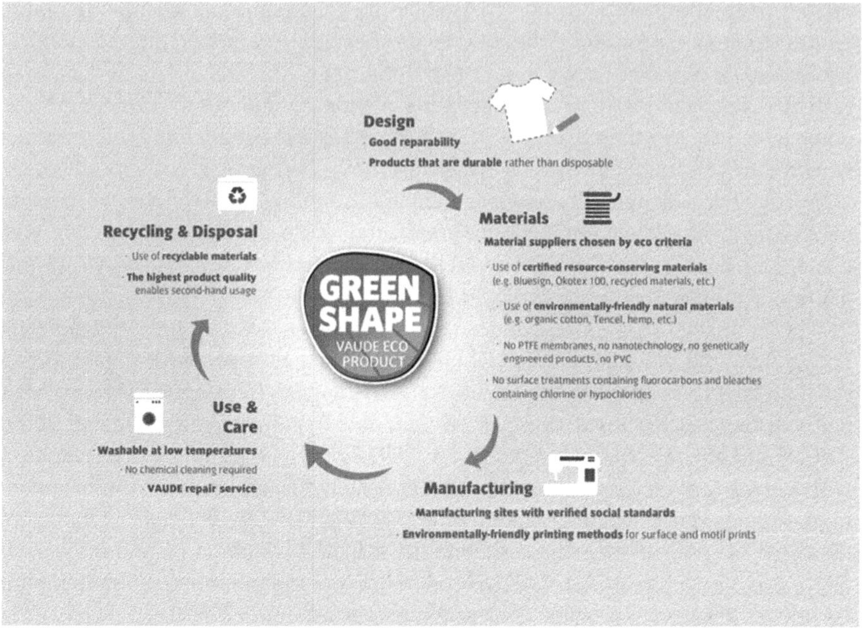

Fig. 4.8 VAUDE green shape components (VAUDE 2017b)

72% (VAUDE 2015). These measures are part of VAUDE's green shape program, which is shown in Fig. 4.8.

The green shape program covers every aspect of the product life cycle from production to recycling and disposal. Moreover, the green shape certification also guarantees ethical work conditions and fair wages, as monitored by independent institutions such as *Fair Wear*. The overall winter collection in 2016 was based on this certification by 60%, whereas the summer collection for 2017 reaches 54% (VAUDE 2016b).

VAUDE established a repair service for its products in order to decrease the number of discarded products. This service is designed to prolong a product's lifecycle and save resources. Moreover, it aims at raising the customer's awareness regarding sustainable consumption.

Since 2014, VAUDE has published its environmental activities according to the *Global Reporting Index*, which also encompasses economic and social performance in a transparent way. Sustainability in its social dimension is another important element for VAUDE. In 2015, the company was named a *leader* in its industry by the Fair Wear Foundation with regard to VAUDE's commitment to fair working conditions worldwide (VAUDE 2016a). Workers employed in Chinese production sites receive average pay which lies 50% above the minimum wage and in Vietnam they receive 115% of the minimum wage. Encouraging sustainable behavior among customers is another aspect of VAUDE's sustainability strategy. Developing consciousness among customers for sustainable consumption is an important measure for educating society about the negative impacts linked to non-sustainable products and production methods (VAUDE 2014).

4.2.3 Higher Purpose

VAUDE's CEO, Antje von Dewitz (2016), defines the higher purpose of VAUDE as the company's *inherent commitment for a livable world.* This commitment translates into social, ecological, and economic dimensions and aims at creating a common good for all. Hence, VAUDE recognizes its inherent responsibility to humanity and nature (VAUDE 2016e). Antje von Dewitz emphasizes that the core of acting consciously at VAUDE is expressed by encouraging a high-trust culture, an integrative stakeholder approach, and a positive conception of human beings.

4.2.4 Stakeholder Integration

Von Dewitz (2016) views a systemic integration of stakeholder needs as a core element of VAUDE which is primarily aligned by its focus on sustainability. Encouraging collaborative communication and trust forms the basis of mutual benefit. Understanding stakeholder perspectives and needs marks the beginning of this process and has to be considered in the entire system of stakeholders, including external organizations such as suppliers, NGOs and distributors. Joint agreements are also aligned to the pursuit of VAUDE's purpose and mission. Antje von Dewitz stresses that stakeholder interests can be integrated without applying a trade-off mindset. Sustainability can only be achieved together and thus requires the contribution of all stakeholders involved. VAUDE is focused on establishing long-term relationships with its stakeholders. For example, around 70% of the producers involved have been partners with the company for five or more years (VAUDE 2014). Promoting further education is another important goal for VAUDE with regard to its employees. VAUDE established the *VAUDE Academy*, which offers events and training about a broad range on topics such as company-specific knowledge transfer, insights about market developments, and new products. In addition, the academy provides courses about the values of the company and its understanding of sustainability as well as yoga classes, mountain bike tours, and further offers which contribute to employee well-being (VAUDE 2016l). The company documents the utilization of the academy's offers in order to conceptualize the program for the following year. The program is available to every member of the organization and can be accessed via the company's social intranet (VAUDE 2016d). With regard to the economic dimension, VAUDE (2014) emphasizes that the creation of a common good is a goal that is superior to focusing on profits as an end. As a result, most of its profits are reinvested in the company to strengthen its stability and equity ratio. In general, VAUDE classifies its stakeholders on a three-

Table 4.2 Stakeholder groups at VAUDE (authors' table)

Stakeholder	Category	Core interests	Inclusion	Priority
Employees	Own Company	Trust	Dialog	1
Shareholder	Own Company	Consistent Reporting	Regular Reporting	1
Subsidiaries	Own Company	Stability	Coordination	1
Suppliers	Suppliers	Procurement	Communication	1
Producers	Suppliers	Partnership	Regular Meetings	1
Banks	Creditors	Risk Management	Regular Reporting	1
Customers	Customers	Integrity	Direct Contact	1
Retailers	Customers	Good Image	Benchmark Studies	1
NGOs	Social Institutions	Transparency	Conferences	2
Administrative	Authority	Compliance	Consultations	2
Certifications	Audit	Credibility	Partnerships	2
Federations	Federations	Knowledge Exchange	Collaboration	2
Municipality	State	Environmentalism	Work Group	2
Government	State	Biodiversity	Regular Meetings	2
Athletes	Sponsoring	Prominence	Events	2
Institutions	Society	Initiator	Direct Contact	3
Media	Media	Transparency	Interviews	3
Universities	Society	Progression	Direct Contact	3
Charities	Society	Support	Direct Contact	3

point scale in terms of priority, the relating category, their core interests, and their inclusion. Table 4.2 shows the network of stakeholders approached according to VAUDE (2016e). The columns *core interests* and *inclusion* each contain only one item for reasons of clarity.

VAUDE values the communities in which it operates and hence supports different regional needs. In 2006, the public swimming pool of Obereisenbach needed an investment in order to continue its operations. VAUDE invested in this facility because it was an important factor for the community's well-being (VAUDE 2014). Another of VAUDE's essential regional investments was the opening of a child care center as a public-private-partnership with the state of Baden-Wuerttemberg. Fueglistaller et al. (2004) note that this engagement meets business and social needs. VAUDE's headquarters are not located near any large cities or universities, so the company has to focus on personnel retention and employer attractiveness. Family has always had a high priority at VAUDE. The childcare offer is a part of VAUDE's strategy to balance family and work among its members. This facility was also needed in order to cope with the above average family size of VAUDE employees (VAUDE 2016f). Nevertheless, many people were skeptical in the beginning about a company getting involved in such an area (von Dewitz 2016). This facility is available to parents at VAUDE as well as to other families in the community. VAUDE also benefits from these measures by reducing costs for retention, employee recruiting, lower absenteeism, and higher levels of employee engagement (Fueglistaller et al. 2004). Fitzner (2009) notes that

these measures show how various stakeholder groups are integrated at VAUDE. In February 2017, VAUDE was honored with the A.T. Kearny Family Award for its ongoing efforts in establishing a family inclusive work environment (VAUDE 2017c). Backhaus-Maul et al. (2008) add that responsible behavior regarding employee well-being is a part of VAUDE's ethical and responsible behavior and reflects the orientation to sustaining its business environment.

Creating consciousness of environmental and social demands is also fostered by cooperation with other kindergartens, environmental associations such as the *Deutscher Alpenverein*, and other partners such as suppliers. With regard to creating a higher sense of consciousness, Antje von Dewitz also identifies a change needed in higher education. According to von Dewitz, ethical and sustainable management should be taught in core rather than elective courses in universities so that the complex system of various stakeholder groups and the need to integrate them in a beneficial manner can be understood. Therefore, von Dewitz also views the role of VAUDE as a pioneer for change in the economic environment and hence as a signal for others. Demonstrating these efforts in a transparent way was also a key influencing factor for VAUDE to become a part of the Economy for the Common Good in 2015. Antje von Dewitz adds that the Economy for the Common Good is a movement which fosters the idea of sustainability and rewards companies for contributing to all stakeholders and not just investors. Moreover, the concept improves sensitization in business and society to fairness and balance of economic, social, and ecologic needs. VAUDE achieves 502 points and demonstrates above-average ratings in the areas of ecological and social product design, reduction of environmental effects, and enhancement of the industry's ecological and social standard (VAUDE 2016g). VAUDE's Economy for the Common Good Matrix is clarified in Fig. 4.9.

4.2.5 Conscious Leadership

Von Dewitz (2016) acknowledges that the leadership style was predominantly patriarchal in the past. Nevertheless, there have been many elements of the nowadays high-trust culture. She notes that the industry can be described in general as collegial and respectful, which also translates to the leadership style. As part of her dissertation about Talent Relationship Management, Antje von Dewitz initiated a transformation of the leadership mentality at VAUDE. This change process visualized areas of improvement and strongly focused on the individuality of every leader, based on a personality model with reference to the DISG approach. This approach uses colors in order to describe different leadership personalities and matches them with personal characteristics. For example, red symbolizes elements such as dynamic, powerful, and determining. Von Dewitz states that these traits were very dominant in the past at VAUDE. Therefore, it was rewarding to explore other traits as well to leverage from different personalities and perspectives on a base of equality. Hence, fostering this understanding and enabling others to lead

CERTIFICATE : AUDIT

COMMON GOOD BALANCE 2013 — for VAUDE SPORT GMBH & CO KG — Auditor Dr. Manfred Kofranek

ECONOMY FOR THE COMMON GOOD
An economic model for the future

VALUE / STAKEHOLDER	Human dignity	Solidarity	Ecological Sustainability	Social Justice	Democratic Co-determination & Transparency
A) Suppliers	A1: Ethical Supply Management 60 %				
B) Investors	B1: Ethical Financial Management 30 %				
C) Staff Incl. Owners	C1: Workplace quality and affirmative action 50 %	C2: Just distribution of labor 40 %	C3: Promotion of environmentally friendly behavior of employees 60 %	C4: Just income distribution 20 %	C5: Corporate democracy and transparency 20 %
D) Customers / Products / Services / Business Partners	D1: Ethical customer relations 40 %	D2: Cooperation with businesses in same field 30 %	D3: Ecological design of products and services 70 %	D4: Socially oriented design of products and services 10 %	D5: Raising social and ecological standards 60 %
E) Social Environment: Region, sovereign, future generation, global fellow human beings, animals and plants	E1: Value and social impact of products and services 50 %	E2: Contribution to the local community 70 %	E3: Reduction of environmental impact 80 %	E4: Investing profits for the Common Good 90 %	E5: Social transparency and co-determination 60 %
Negative Criteria	Violation of ILO norms (international labor standards) / human rights 0 Products detrimental to human dignity and human rights (e.g. landmines, nuclear power, GMO's) 0 Outsourcing to or cooperation with companies which violate human dignity 0	Hostile takeover 0 Blocking patents 0 Dumping Prices 0	Massive environmental pollution 0 Gross violation of environmental standards 0 Planned obsolescence (short lifespan of products) 0	Unequal pay for women and men 0 Job cuts or moving jobs overseas despite having made a profit 0 Subsidiaries in tax havens 0 Equity yield rate >10 % 0	Non-disclosure of subsidiaries 0 Prohibition of a works council 0 Non-disclosure of payments to lobbyists 0 Excessive income inequality within a business 0

This certification confirms the audit of the common good report. The certificate refers to the common good matrix 4.1. Detailed information concerning the matrix, the indicators and the audit system can be found on www.economy-for-the-common-good.org

Certification valid 31.01.2017 — END BALANCE 502

Fig. 4.9 Economy for the common good matrix of VAUDE (2016h)

with this mindset is a key responsibility for the CEO. Changing the notion of leadership was also a necessity for VAUDE to cope with the dynamic changes of the industry and manage the complex corporate structure. In addition, clear communication and collaboration across functions and departments in combination with a high-trust culture symbolize core success factors for VAUDE in order to innovate continually. Antje von Dewitz adds that inter-human relations can be a source of conflict but also represent a great potential for achieving the company's goals as a community. Therefore, a patriarchal leadership style would not meet the needs of the employees today at VAUDE as there is an increasing demand for autonomy and purpose combined with an equal approach to leadership. Although there are employees who might favor a patriarchal style, VAUDE is convinced that the new leadership model is a basis for long-term success. Hence, leading with integrity and openness to different perspectives is both expected and supported for leaders in the organization. In March 2017, the company received the Soul@Work Award for putting its culture of trust into practice and for its outstanding focus on employee well-being (VAUDE 2017d). VAUDE offers various methods for developing this leadership style by offering coaching and seminars where leaders and future leaders learn how to enable others to increase their self-efficacy and active contribution. To strengthen this understanding, VAUDE selected a group of employees as leaders who embody this perspective of leadership based on their high-trust cultures and who act as multipliers (von Dewitz 2016).

4.2.6 Conscious Culture and Management

VAUDE is a value-driven organization which bases its teamwork on trust among the organization's members. Trusting employees is perceived as a necessity for people to reach their full potential and contribute to the company's performance. VAUDE's trust culture is based on a positive image of employees regarding their willingness to assume responsibility and contribute their work to the organization (VAUDE 2016i). In 2014, the University St. Gallen started a research project at VAUDE and came to the conclusion that the company increasingly developed a culture of trust (Weibel and Schafheitle 2015). In conjunction with this research project, VAUDE and the University St. Gallen conduct regular surveys on the leaders' and employees' satisfaction and collaboration. The results are used to identify potential areas of conflict and improvement and then to work on them.

In addition, the company takes a close look at indicators like employee turnover rate and the sick leave rate to assess the current status. As part of the transformation of VAUDE, an internal culture committee was established to reinforce the culture. This group consists of 50 managers and 15 employees who act as *Kulturträger* (von Dewitz 2016). With reference to this, she perceives her role as a guardian of the company's culture.

4.2.7 Company Performance

VAUDE emphasizes that the company is not obligated to publish financial figures and decided not to state absolute numbers for reasons of competition (VAUDE 2016d). VAUDE is a key player in the German outdoor market, as can be seen in its revenues. Figure 4.10 shows the market distribution according to VAUDE (2017e).

Nevertheless, the company publishes a relative development of its revenue in comparison to the European Outdoor Group, which represents this industry in Europe. Figure 4.11 highlights VAUDE's revenue development between 2011 and 2015 (VAUDE 2017e).

As presented, it was not possible to maintain the strong revenue development in 2011 through the year 2015. VAUDE (2016j) states that besides the decreasing development of revenues, the company could constantly realize an above-industry average growth. Figure 4.12 outlines the growth rate development in comparison to the European Outdoor Group (EOG) (VAUDE 2016e).

The company has four limited partners, who are all family members. Profits which are generated beyond profit withdrawals by the four limited partners are used entirely for internal investments and to increase the company's equity ratio to foster financial independency and stability. The equity ratio increased from 39.2% in 2014 to 47.7% in 2015 while the relative liabilities decreased from 69.2% in 2014 to 67.07% in 2015 (VAUDE 2016d). With regard to the employee turnover rate, VAUDE achieved a rate of 6.1% in 2015 (4.7% in 2014) and aims at stabilizing the rate around 5% (VAUDE 2016k). In 2013, the average fluctuation rate was around 14% in Germany (Hay Group 2013). Customer satisfaction represents a key success factor and a high priority for VAUDE. Therefore, VAUDE aims at intensifying the dialogue with both end customers and its dealership network to increase satisfaction among both groups. With reference to this, the company carefully evaluates complaints in order to minimize negative impacts. For this reason, VAUDE implemented a customer service portal called *Zendesk* where customers can share their issues with the company and thus contribute to areas such as

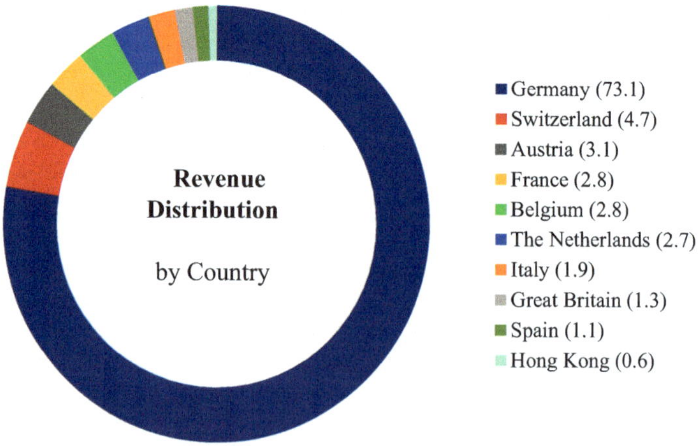

Fig. 4.10 VAUDE—revenue distribution by country (authors' diagram)

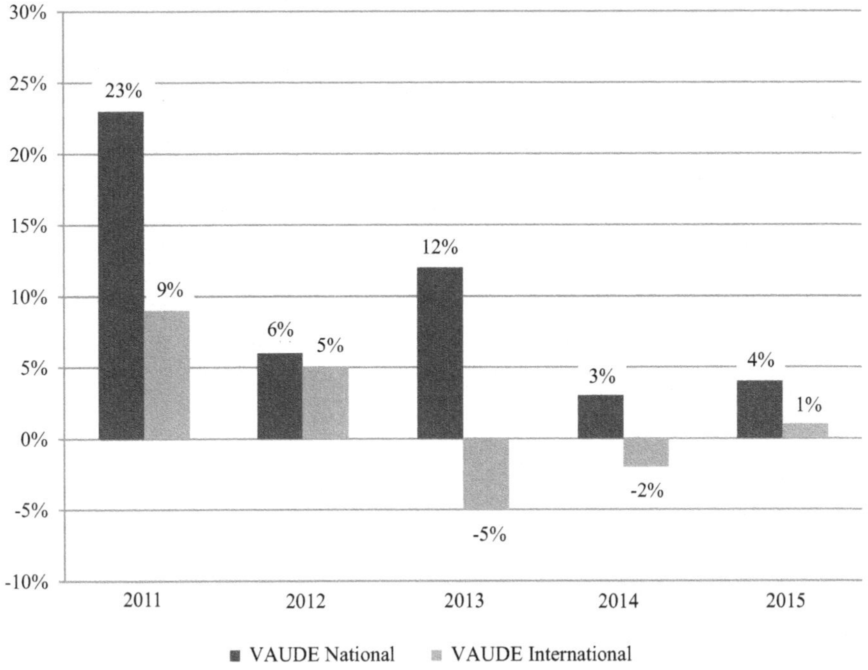

Fig. 4.11 VAUDE's revenue development from 2011 to 2015 (author's diagram)

product development and quality management. VAUDE evaluates the information and calculates a satisfaction score. In 2015, VAUDE achieved a score of 77.7% (63.7% in 2014) based on the evaluation of 9000 inquiries (6300 in 2014) of which 41.9% (39.2% in 2014) have been rated by customers (VAUDE 2016j). Since VAUDE does not sell to end customers directly, it focuses on achieving good relationships with its dealer network. In a benchmark study conducted by Benchmark Excellence, VAUDE was able to defend its leading position from 2013 to 2015.

This benchmark analyzes the satisfaction of dealers with outdoor companies on the dimensions *customer service, support for business development, logistics and supply chain, dealer marketing, portfolio development, and personnel* (VAUDE 2016h, Table 4.3).

4.2.8 Conclusion

VAUDE is a company that is strongly driven by its values and its orientation to achieving sustainability among its activities. It focuses on inspiring others to follow and collaborates with different institutions, NGOs and its partners such as suppliers and producers. Von Dewitz (2016) defines the higher purpose of VAUDE as an inherent commitment for a livable world. In May 2017, she received the Order of Merit of the state of Baden-Wuerttemberg, especially for her environmentally

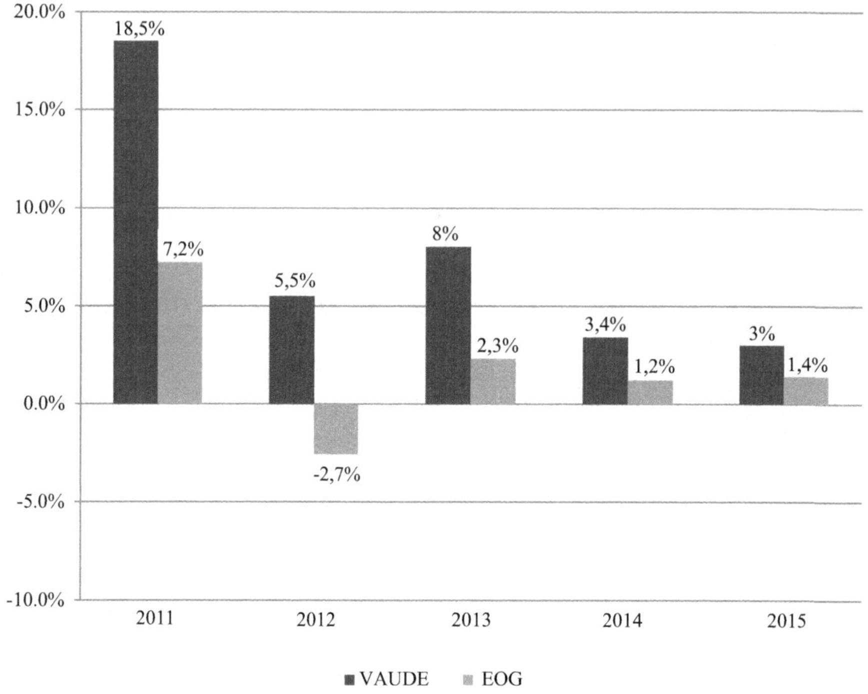

Fig. 4.12 Growth rate development of VAUDE and the EOG from 2011 to 2015 (author's diagram)

Table 4.3 Benchmark excellence scores 2013 & 2015 (authors' table)

Company	Overall score (%)	Rank 2015	Rank 2013
VAUDE	73	1	1
Mammut	72	2	3
Schöffel	71	3	4
Meindl	71	4	4
Jack Wolfskin	70	5	8
Lowa	70	6	2
Bergans	65	7	9
Fjäll Räven	65	8	6
Salewa	63	9	9
Haglöfs	62	10	12
Hanwag	62	11	7
Marmot	62	12	9
The North Face	61	13	13

conscious and responsible acts as a business woman. Regarding the types of higher purposes by Mourkogiannis et al. (2008) and Mackey and Sisodia (2014) in Sect. 2.6.1, VAUDE relates to *the heroic* dimension as it centers its purpose towards the global contribution of its business in terms of making the world a better place by acting consciously and sustainably. Having a strong orientation on ecologic impacts is emphasized by Laloux (2014) as an essential trait of conscious businesses. VAUDE strongly focuses on contributing to a common good on an ecologic, social, and economic dimension. The company demonstrates a focus on humanity and nature by its sustainable business activities, ethical work conditions in Germany and abroad, support for communities, the implementation of strict sustainability standards and the fostering of consciousness for these topics among its stakeholders. Behaving in a social and environmentally conscious manner represents the core orientation at VAUDE. Thus the company recognizes that its purpose can only be achieved in collaboration with its stakeholders. The company focuses its vision on a long-term timeframe and rejects the idea of trade-off thinking. This rejection has been postulated as an essential element in the stakeholder integration of conscious businesses by Sisodia et al. (2014) and Freeman et al. (2007). Transparency plays a significant role for the company with reference to measuring its impact on stakeholders.

This is why VAUDE decided to apply strict and comprehensible reporting standards such as the GRI index, the Economy for the Common Good, Fair Wear, bluesign, and other certification standards. Moreover, VAUDE established its own standard regarding its product portfolio which goes beyond the existing frameworks in terms of sustainability. VAUDE is convinced that the success of its activities depends on the value of trust among its members. Although the company already had elements of a high-trust culture before Antje von Dewitz took over the management, VAUDE intensified its efforts in promoting this direction. This is supported by an internal committee which acts as a multiplier for living and sustaining the culture. A high-trust culture has been identified as an essential element of conscious businesses by Sisodia et al. (2014). Von Dewitz considers that it is essential to acknowledge the individual potential of each and every employee. This is why she initiated a transformation of the leadership style from patriarchal to a new form which focuses on autonomy and self-efficacy among leaders and employees alike. Leaders are seen by employees as initiators and supervisors of this mindset rather than as leaders whose leadership is based on advice and limitations. Sisodia et al. (2014), Mackey and Sisodia (2014), and Bill George (2014) regard the orientation to autonomy and self-reliance as a distinct characteristic of a conscious approach to leadership. With reference to the TACTILE elements of conscious cultures, VAUDE fulfills the criteria based on the findings because it fosters transparency, trust, and accountability as well as loyal treatment of stakeholders, adherence to truthful, socially acceptable, fair, and ethical behavior in support of human needs. VAUDE does not pursue profits for the sake of profits, but prioritizes the creation of a common good as guiding principle of its business activities. Profits are distributed to a certain extent to the company's limited partners who are all members of the family. Additional profits are reinvested in the company to strengthen its financial stability and independency. With regard to revenue development, VAUDE does not publish absolute financial metrics. However, the company does communicate its relative revenue development, which has been above the industry average for the past five years. Moreover, VAUDE has been honored as leader in the outdoor equipment industry in regard to its dealership network. Concerning its end customers, the company shows an increase in satisfaction based on the evaluation of its customer platform. In comparison to the average fluctuation rate in Germany, VAUDE shows a significantly lower employee turnover rate. However, the average turnover rate encompasses all industries and not just the industry in which VAUDE operates. The findings described relate to the company's superior performance, which is identified as a characteristic of conscious businesses by Mackey and Sisodia (2014). Based on the findings described, VAUDE fulfills a majority of criteria from the research framework and can thus be considered as embodying conscious business principles in Germany.

4.3 Polarstern

Polarstern is the subject of the third case study (Fig. 4.13). The information about conscious business principles is derived from available company information, articles, and an interview with one of its founders, Florian Henle. This case focuses on the implementation of conscious business principles with an emphasis on the company's purpose and vision, its focus on stakeholder management and its sustainability efforts.

4.3.1 Company Development

Polarstern was founded in 2011 as a social business by Jakob Assmann, Florian Henle, and Simon Stadler (Polarstern 2016a). The company was established with a shared vision to transform the world by promoting 100% renewable energy. Polarstern buys renewable energy and biogas and transfers these products to their customers. Hence, Polarstern acts as an intermediary trading partner between producers and customers (Assmann 2016). In 2012, the products of Polarstern were certified by TÜV Nord [TÜV = Official Technical Inspection Agency] and achieved the *Grüner Strom Label* with a gold award from 2013 to 2015 (Polarstern 2016b). In 2013, VAUDE switched to Polarstern in order to support its mission of becoming the most sustainable outdoor fashion manufacturer (Polarstern 2016c). Polarstern was the first company from the energy sector to join the Economy for the Common Good (Polarstern 2016d). It did so to support transparency and social business practices.

Company Profile

Polarstern
Wirklich bessere Energie.

(Polarstern 2017a)

Founded	2011 in Munich (Germany)
Industry	Energy
Headquarter	Munich
CEO	Jakob Assmann, Florian Henle, Simon Stadler
# Employees 2015	19
Revenue 2015	€2.4 million

Fig. 4.13 Polarstern Company profile (authors' diagram)

4.3.2 Sustainability Efforts

Polarstern was the first energy supplier in Germany to offer 100% renewable energy and biogas. Assmann (2016) remarks that there is a large number of energy and gas suppliers in Germany, but many of them who claim to offer renewable energy and biogas do not publish their chain of production, which results in a lack of transparency. At Polarstern, the electricity sold is generated by hydroelectric facilities in Germany and Austria. The production of biogas is based on the use of 100% organic recycling material such as sugar beets processed in Hungary. In order to diminish environmental effects of their products, Polarstern cooperates with many partners to promote measures such as renaturation of rivers. The entire value chain is aligned to the creation of purely ecological products. This benefits the common good and is not primarily designed to earn high profit margins. Polarstern recognizes this orientation as their unique selling proposition in comparison to other energy suppliers. The founders established this company to help people to switch from semi-renewable energy to purely renewable energy and to mitigate the effects of climate change, pollution, and other environmental risks.

Every customer of Polarstern contributes to a change to renewable energy worldwide since the company invests a certain amount of every used kilowatt hour in sustainable energy projects around the globe (Polarstern 2016b). Moreover, Polarstern aims at educating people on sustainable energy consumption by giving background information regarding the energy market and worldwide trends (Polarstern 2016e). In order to decrease energy costs for both landlords and tenants, Polarstern supports the model of *Mieterstrom*. Mieterstrom means that landlords and tenants can profit from using solar panels and communal heating stations to cover a part of their energy demand and combine it with products from energy suppliers. This model results in a higher degree of independency and a decrease of costs as well as a more sustainable way of generating energy. Polarstern offers to buy the energy of landlords which is generated by solar panels and communal heating stations at an above-average industry price. This energy is then sold by Polarstern to the tenants for a below-average price combined with energy and gas products from Polarstern. Hence, landlords are able to profit by earning a return on their investment and tenants profit by paying less for their energy. Polarstern has established three projects in Munich which are using this model successfully (Polarstern 2016g). The German Government recently released a draft bill to support the development of this model to advance a sustainable energy transition by actively including private households (BMWi 2017). In 2017, Polarstern and Panasonic announced a strategic cooperation regarding thermal heat pumps to advance the energy transformation to an emission-free solution in heating systems. This cooperation bundles the vision of Polarstern and Panasonic in a beneficial manner and offers customers a specialized package (Polarstern 2017b). In addition to the impact in Germany, every customer of Polarstern supports a family in Cambodia with 20 € of the annual fee, which is invested in micro-biogas facilities. The energy production in Cambodia is predominantly based on burned materials such as wood, which results in higher levels of pollution and diminishes the ability to cover CO_2 emissions. Therefore, Polarstern's approach directly benefits families and the environment in Cambodia and also results in lower energy costs for families because they are then able to use organic waste from their agricultural production (Polarstern 2016f). Hence, Polarstern recognizes the importance of both raising awareness in Germany and creating an impact worldwide with regard to supporting renewable energies.

4.3.3 Higher Purpose and Vision

Polarstern aims to make the world a better place by promoting the development of renewable energies worldwide. According to Henle (2016), this orientation and philosophy has been an inherent principle of Polarstern's since it was founded. The company's higher purpose translates into an active way of shaping the energy transformation worldwide. Polarstern wants to create a signal that the change to renewable energies is both beneficial for society and also valuable in a competitive market. Polarstern's vision includes creation of consciousness for an individual's contribution as well as raising awareness for a sustainable and mindful use of energy. In addition, the decarbonization of business and society as well as the reduction of global inequalities regarding clean energy are core aspects of this company. Polarstern believes that the transformation to renewable energies can be significantly supported by raising awareness in society rather than waiting for large energy conglomerates to change their operations (Polarstern 2015).

Nevertheless, Florian Henle adds that Polarstern is able to contribute to a systemic change in large corporations by demonstrating the effectiveness of their business operations. Polarstern's operation is based on achieving its higher purpose and is not driven by economic growth itself. Moreover, Polarstern is a member of the *Purpose 100 Economy*, which lists purpose-driven businesses worldwide (Polarstern 2016g). This business model depends on integrating a social, ecological, and economical yield in a uniform manner. With regard to this orientation, Henle views profitability as a means to operate the business with stability and flexibility and emphasizes that Polarstern pursues this element as a means and not an end in itself.

4.3.4 Stakeholder Integration

Polarstern pursues a collaborative approach in meeting stakeholder needs. Integrating various needs is achieved by clear communication regarding the company's purpose and vision. Therefore, all interests are considered with regard to their impact on an ecological, societal, and economical level. This mindset is valid for internal and external stakeholders (Polarstern 2015). Henle (2016) emphasizes that this approach is based on the premise of fairness, openness, and transparency regarding the communication of business activities. If stakeholder interests deviate from the Polarstern philosophy, management tries to find a mutually beneficial solution. Nevertheless, Polarstern is also willing to end a relationship if the issue entails an insurmountable contrast to the company's philosophy. The principle of fairness is visible regarding the pricing model and the contract liabilities with customers. With reference to that, Florian Henle states that Polarstern does not claim a premium for its products although this would be possible according to their position. Polarstern emphasizes that the price is comparable to the average price for non-renewable energies to offer an affordable product. In addition, the product itself is described in a transparent manner concerning the production and distribution cycle. The information regarding the value of Polarstern is publicly accessible and documents the entire process. Polarstern is supported by private investors as well as by the *Gesellschaft Energie 360 Grad* which is owned by the city of Zurich. Polarstern is financed 70% from outside capital. The Gesellschaft Energie 360 Grad is both a stakeholder and a customer of Polarstern and owns 20% of the company. This partnership involves a mutually beneficial bond because both parties share the same vision and benefit from the partnership in terms of knowledge exchange and financial support. Florian Henle highlights that Polarstern uses the capital in a highly responsible manner because it values the trust and integrity connected with the investments. With reference to this, Henle views the mindset of Polarstern as a clear contrast in comparison to exit-driven start-ups. Polarstern focuses on being part of an ethical network in terms of its business environment. Since it was founded, the company decided to establish a long-term partnership with the GLS Bank and the Triodos Bank in order to support their ethical and sustainable financial model. Besides creating a common good for society, Polarstern also promotes creation of good working conditions. Shaping tasks in a purposeful and challenging manner and establishing a healthy work-life balance and fair wages belong to its core philosophy. Openness to change, a strong sense of participation and responsibility, integrity, and respectful communication are both supported and expected at Polarstern.

The company aims at providing a platform for personal and professional development and supports this orientation by personal supervision, workshops, and regular meetings (Polarstern 2015). Florian Henle adds that this philosophy is lived by the management team as well as by employees and is based on equality regarding the importance of individual contributions. The employment strategy of Polarstern is focused on internal development and promotion. Half of the personnel have been promoted from an intern level to full-time employee. The personal planning strategy is transparent and accessible by every member of the organization. Furthermore, Polarstern offers flexible working hours according to the individual needs of employees. Home office solutions are also offered under the premise that employees sometimes spend more time in the office than at home. An additional vacation day is granted once a year for charitable activities. The physical and psychological well-being of employees is a high priority for the company. Hence, Polarstern offers a *freeletics* training program twice a week and intensive yoga classes twice per month. With regard to fair employee payment, the variance between the highest and lowest income equals two to one. Although Polarstern is a social business, the wages are comparable to the industry average. Moreover, there is no gap between wages of male and female employees on all levels. Establishing collaborative partnerships among stakeholders also relates to competitors and other parties in the value chain. The company supports others with regard to their development in the sector of renewable energies because Polarstern perceives it as an essential way to learn from each other and inspire others for their path (Polarstern 2015). This commitment is especially manifested with companies that are part of the label *Grüner Strom*. Hence, Polarstern acts in favor of its purpose and does not condemn activities of competitors (Henle 2016). Collaborative marketing constitutes another principle of Polarstern's which is carried out with other companies. Polarstern has established partnerships with companies such as Patagonia, VAUDE, Coffee Circle, Viva con Agua and others. These partnerships aim at multiplying relevant content which unites these companies on different media channels. Polarstern supports the well-being of the communities in which it operates and hence supports different initiatives. In 2016, the company organized a cleanup project for the Isar river in Munich in cooperation with other partners such as Patagonia. This project aimed at raising awareness for local environmental problems which are related to waste (Polarstern 2016i). As a result of this orientation, the company supports the Economy of the Common Good as the first energy supplier (Polarstern 2015). Polarstern is committed to creation of a common good and thus decided to become a part of the Economy for the Common Good. The company has been evaluated by a peer group which consists of *Impact Hub*, *Kartoffelkombinat*, and *Talents4Good* (Polarstern 2016h). Florian Henle remarks that the Economy for the Common Good is a valid instrument for Polarstern to communicate its efforts in a transparent manner. Polarstern achieved a total score of 652 points and did not display any of the negative criteria. The Economy for the Common Good Matrix of Polarstern is summarized in Fig. 4.14.

4.3.5 Conscious Culture, Management, and Leadership

Establishing a conscious approach to organizational culture and management is a high priority for Polarstern. Florian Henle adds that this development is mainly based on the principle of equality and personal responsibility, to establish an environment where people can benefit from each other by sharing perspectives. With regard to its business activities, Polarstern emphasizes the importance of creating awareness for environmental impacts among its members. This is also supported by the exchange of knowledge and ideas with environmental institutions such as Sea Shepard, Surf Rider Foundation and sustainable companies such as Kartoffelkombinat, Bleed and other partners. With reference to this perspective, Henle states that the management team considers themselves as role models for other members of the organization. Hence, the mindset of acting in a conscious and

TESTAT : PEER EVALUIERUNG

GEMEINWOHL-BILANZ 2014 — für Polarstern GmbH — BegleiterIn Nikolaus Teixeira

GEMEINWOHL **ÖKONOMIE** Ein Wirtschaftsmodell mit Zukunft

WERT / BERÜHRUNGSGRUPPE	Menschenwürde	Solidarität	Ökologische Nachhaltigkeit	Soziale Gerechtigkeit	Demokratische Mitbestimmung & Transparenz
A) LieferantInnen	A1: Ethisches Beschaffungsmanagement 60 %				
B) GeldgeberInnen	B1: Ethisches Finanzmanagement 60%				
C) MitarbeiterInnen inklusive EigentümerInnen	C1: Arbeitsplatzqualität und Gleichstellung 50 %	C2: Gerechte Verteilung der Erwerbsarbeit 40 %	C3: Förderung ökologischen Verhaltens der MitarbeiterInnen 60 %	C4: Gerechte Verteilung des Einkommens 60 %	C5: Innerbetriebliche Demokratie und Transparenz 40 %
D) KundInnen / Produkte / Dienstleistungen / Mitunternehmen	D1: Ethische Kundenbeziehung 70 %	D2: Solidarität mit Mitunternehmen 60%	D3: Ökologische Gestaltung der Produkte und Dienstleistungen 90 %	D4: Soziale Gestaltung der Produkte und Dienstleistungen 70 %	D5: Erhöhung der sozialen und ökologischen Branchenstandards 70 %
E) Gesellschaftliches Umfeld: Region, Souverän, zukünftige Generationen, Zivilgesellschaft, Mitmenschen und Natur	E1: Sinn und gesellschaftliche Wirkung der Produkte / DL 90%	E2: Beitrag zum Gemeinwesen 80 %	E3: Reduktion ökologischer Auswirkungen 70 %	E4: Gemeinwohlorientierte Gewinnverteilung 70 %	E5: Gesellschaftliche Transparenz und Mitbestimmung 70%
Negativ-Kriterien	Verletzung der ILO-Arbeitsnormen/Menschenrechte 0 Menschenunwürdige Produkte, z.B. Tretminen, Atomstrom, GMO 0 Beschaffung bei / Kooperation mit Unternehmen, welche die Menschenwürde verletzen 0	Feindliche Übernahme 0 Sperrpatente 0 Dumpingpreise 0	Illegitime Umweltbelastungen 0 Verstöße gegen Umweltauflagen 0 Geplante Obsoleszenz (kurze Lebensdauer der Produkte) 0	Arbeitsrechtliches Fehlverhalten seitens des Unternehmens 0 Arbeitsplatzabbau oder Standort-verlagerung bei Gewinn 0 Umgehung der Steuerpflicht 0 Unangemessene Verzinsung für nicht mitarbeitenden Gesellschafter 0	Nichtoffenlegung aller Beteiligungen und Töchter 0 Verhinderung eines Betriebsrats 0 Nichtoffenlegung aller Finanzflüsse an Lobbies / Eintragung in das EU-Lobbyregister 0 Exzessive Einkommensspreizung 0

Mit diesem Testat wird die Peer-Evaluierung des Gemeinwohl-Berichtes bestätigt. Das Testat bezieht sich auf die Gemeinwohl-Matrix 4.1. Nähere Informationen zur Matrix, den Indikatoren und dem Audit-System finden Sie auf www.gemeinwohl-oekonomie.org

Testat gültig bis 31.12.2017 — BILANZSUMME 652

Fig. 4.14 Economy for the common good matrix of Polarstern (Polarstern 2016h)

sustainable manner has to be demonstrated by the founders in order to pass it on to their employees and other stakeholders. Polarstern believes that their purpose needs to be visible in all interactions with stakeholders to support trust and integrity regarding their business model (Polarstern 2015). Mutual respect, transparency, and trust are core characteristics of the Polarstern approach of culture and leadership.

4.3.6 Company Performance

Polarstern is a social business and hence does not place a primary emphasis on accumulating profits. In the fiscal year 2015/2016, the company did not distribute any profits to its investors. Polarstern's financial philosophy is focused on reinvesting into its equity ratio, infrastructure and company structure to further strengthen the business model. The company emphasizes that the lenders of capital share the same understanding and hence do not expect high profits. In 2014, Polarstern generated revenues of 2.4 million €. The company operates with low profit margins and thus concentrates on gross margin per customer. Polarstern states that they invested a double digit percentage of the average gross margin per customer in the development of projects in Cambodia. Moreover, employees benefit from the reinvestment strategy with a yearly bonus payment (Polarstern 2015). In terms of customer satisfaction Polarstern achieved a ratio of 99.41% positive reviews out of 1183 ratings at the independent rating portal eKomi (eKomi 2016).

4.3.7 Conclusion

Polarstern is a purpose-driven organization with a strong focus on creating a common good on social and environmental dimensions. The company aims at making the world a better place by promoting renewable energy solutions. With regard to the classification by Mourkogiannis et al. (2008) and Mackey and Sisodia (2014), Polarstern relates to the ideal of *the heroic*. Henle (2016) clarifies that the purpose constitutes the inherent nature of the company which is also visible in its efforts to support families in Cambodia with renewable energy solutions. Kofman (2006) and Laloux (2014) view the fulfillment of a common good, based on a human-centric approach, as an essential criterion of conscious businesses which is present at Polarstern. The involvement of stakeholders is an important factor for Polarstern for promoting consciousness regarding various areas of sustainable living. This translates into educational measures such as the *Isar Cleanup Project*, guidelines for sustainable consumption, and partnerships with other companies that share the same perspective on environmental and social sustainability.

Integrating stakeholders in a conscious, respectful, and beneficial way is an inherent aspect of Polarstern. The company's stakeholder management approach includes customers, employees, investors, the environment, society and competitors. Polarstern believes that its vision is best achieved if it is able to mobilize all partners, including competitors. Changing the business environment to clean and renewable energies through a conscious change in behavior and knowledge exchange is perceived as a guiding principle at Polarstern. As a result, the company rejects the idea of trade-off thinking regarding their stakeholders. Sisodia et al. (2014) and Freeman et al. (2007) see this rejection as an inherent element of conscious businesses. According to the findings in Sect. 4.3.3, Polarstern embodies this perspective in its philosophy and business activities. With reference to the leadership and management approach, Polarstern views its employees as equal partners and values individuality and autonomy rather than relying on guidance by supervisors. Moreover, the founders view authenticity and integrity as a daily necessity for their leadership approach. These characteristics are mentioned by Sisodia et al. (2014) and George (2014) as directly connected to a conscious approach to leadership. Transparency, trust, and accountability are essential values for Polarstern in regard to its business model and the company's culture. Additionally, the findings indicate that Polarstern promotes loyal treatment of stakeholders, as well as adherence to truthful, socially acceptable, fair, and ethical behavior in support of human needs. According to the results from the interview, the company also rejects hierarchical systems and centralized decision making processes. Therefore, Polarstern fulfills the criteria of TACTILE cultures which are postulated by Mackey and Sisodia (2014). With reference to the company's performance, Polarstern is not a profit-oriented business. Florian Henle said that the company does not aim at adding a price premium or using other strategies to increase the profit beyond the amount needed for financial stability. It is not clear whether the company operates successfully on the premise of financial factors such as profit margins, revenue development and other metrics. Nevertheless, Polarstern has been achieving a high level of customer satisfaction, which is a characteristic of conscious businesses according to Mackey and Sisodia (2014). Overall, Polarstern fulfills a majority of conscious business characteristics and thus can be considered as embodying these principles in Germany.

4.4 dm drogerie markt

dm drogerie markt is the subject of the fourth case study (Fig. 4.15). The information regarding conscious business principles is derived from available company information, books, and an interview with its CEO, Erich Harsch. This case focuses on the implementation of conscious business principles with an emphasis on the company's dialogical leadership model, its focus on empowerment and decentralization and its anthroposophical core.

Company Profile

(dm drogerie markt 2017a)

Founded	1973 in Karlsruhe (Germany)
Industry	Retail
Headquarter	Karlsruhe
CEO	Erich Harsch
# Employees 2016	56,537
Revenue 2016	€9.708 billion

Fig. 4.15 dm drogerie markt Company profile (authors' diagram)

4.4.1 Company Development

dm drogerie markt was founded by Götz W. Werner in Karlsruhe in 1973. It offers a wide range of drogistic products, from cosmetics to pet food, photography, children's clothing and organic food to household and cleaning products. This approach includes a variety of private brands such as ebelin, dmBio, alverde and Babylove. In German-speaking countries, the market is characterized by other market players such as Rossmann and Müller. At the international level, retailers such as Walgreens Boots Alliance and RiteAid offer comparable product lines. After dm's successful introduction in Germany, Werner opened the first dm store in Austria three years later. As the concept developed quickly, dm established various programs for employees to strengthen their personal and professional capabilities. This focus has been an integral part of dm in order to maintain a human and customer-centric orientation. After German reunification in 1990, dm expanded to the eastern parts of Germany as well as to Hungary and the Czech Republic. On the company's 25th anniversary, dm celebrated its success by hiring 700 new apprentices. In addition, dm donated 4.3 million Deutschmarks to social institutions because Götz Werner identified support for people and communities as an integral part of its business activities (dm drogerie markt 2016a). dm has ranked among the 500 most valuable companies in Germany since 1996 (Pullig 2016). In 2001, the company was appointed as industry leader in regard to customer satisfaction and has maintained this position ever since. Five years later, dm started its cooperation with the Alanus University of Applied Sciences to create a dual education system. In 2008, dm opened its thousandth store in Germany. Moreover, the company established the initiative *be a futurist* in cooperation with the UNESCO to support 1000 sustainability projects which tackle the challenges of the future. In the same year, Erich Harsch was appointed as new managing director and Götz Werner joined the supervisory board. Two years later, Werner transferred his company stake to the charitable *dm-Werner* foundation. Since 2012, all dm stores, logistic centers, and the headquarters have been operated exclusively with electricity from renewable energies.

Since 2016, dm has also been engaged in a pilot project on e-mobility in logistics in the Cologne delivery area in order to reduce the CO_2 load and to achieve

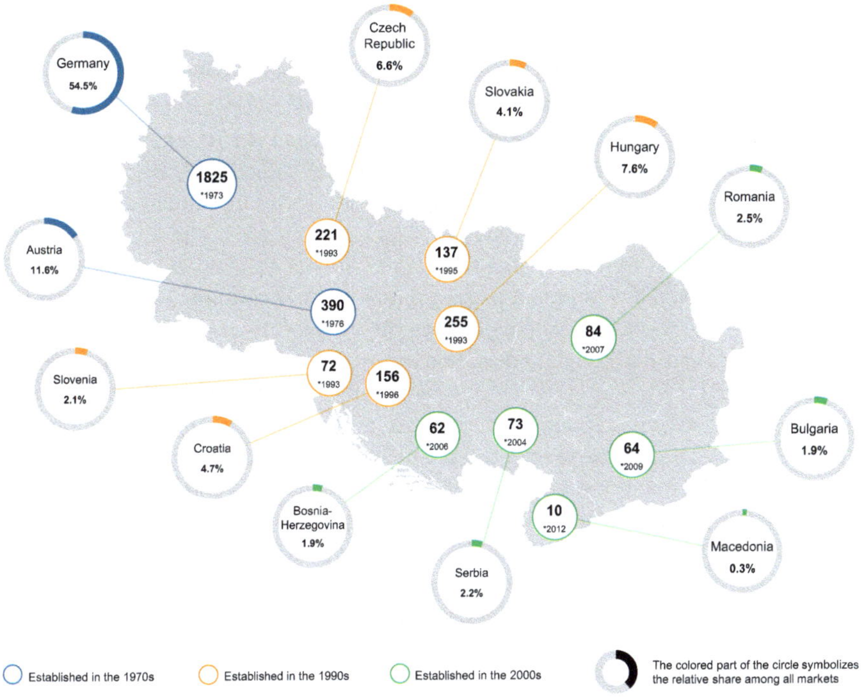

Fig. 4.16 dm markets in Europe (authors' diagram)

emission-free and environmentally compatible solutions in this area as well. In 2014, Götz Werner received a lifetime achievement award for his effort in creating dm as a business based on appreciation of human development, purpose, and responsibility. The company has been constantly expanding its business activities in Europe with over 3000 stores in 12 European countries as of 2016 (dm drogerie markt 2016a). The geographical distribution in Europe is clarified within Fig. 4.16.

dm also offers a selection of its products via an online store and in 2017 it started to sell a range of articles in China through the online platform Tmall. Chinese customers can order a selection of dm's private brands as well as a milk powder of a German producer from an authorized reseller, which is very popular in China. This is the first time dm has distributed outside Europe (dm drogerie markt 2017c).

4.4.2 Anthroposophical Philosophy

The founder, Götz Werner, perceives the nature of business as being driven by satisfying human needs. Hence, he views businesses as relying on people and not vice versa. Werner emphasizes the importance of exploring human potential in every member of the organization and the people who are affected by its operations.

This perspective formed his understanding in setting the core of dm drogerie markt on anthroposophy (Werner 2007).

Werner reframed the quote by Johann Wolfgang von Goethe from *hier bin ich Mensch, hier darf ich's sein* to *hier bin ich Mensch, hier kauf ich ein [Here is where I am a human being; here is where I shop.]*. The slogan clearly expresses a human dimension of customers and employees, who are valued as individuals at dm (Werner 2007). With reference to Kofman (2006), fostering humanity and human development is a key characteristic for conscious businesses. The founder believes that unlearning and reflecting on beliefs and knowledge is an essential aspect of fostering individual development in a conscious way. According to this perspective, consciousness arises when people question their perspectives and hence are able to reject unconscious thinking and behavior as a part of a continuous learning process (Werner 2013).

Essentially, Werner recognized a societal shift in values to an increasing importance of individual empowerment and reflected this change at dm (Amann and Khan 2011). Although the anthroposophical approach has been existent since dm was founded, its organizational implementation mainly developed in the 1980s as part of a long-term business transformation. Hierarchical levels have been reduced to empower employees with more individual responsibility and give them the needed space for experiencing inner development (Pullig 2016). Moreover, this transformation led to a focus on collective value creation, a high level of agility and creativity, entrepreneurial thinking, and a fundamental shift from responding to customer need rather than to customer demand (Pfläging and Selders 2012). Werner views businesses as *living organizations* and portrays the company as a human entity. Hence according to his belief, there are three core aspects of living organizations which have to be consciously embraced at dm. The first driver is *freedom*, which relates to individual initiative, empowerment, self-reliance, individual decisions, and voluntariness. Moreover, *equality* represents a prerequisite for valuable teamwork and values the rules for interaction. The third aspect is referred to as *respect* for the interdependences in interactions and coordination of decision processes. This value is valid both internally and externally. In addition, dm focuses on supporting self-learning competences for every member of the organization (Bauer et al. 2004). With regard to its sustainable development, dm perceives this learning approach as essential to consciously reflecting on individual and organizational impacts and hence adapts individual behavior in a sustainable manner (dm drogerie markt 2016c). This element is also a core aspect of the company's apprenticeship program, where people are encouraged to use an individual trial-and-error approach in real-life situations. This procedure helps the program participants to reflect on their abilities to solve problems and learn about their mental models which are active during this process at an individual pace (Bauer et al. 2004). Hence, achieving a sense of self-mentoring with the help of specialized supervisors and mentors is a valuable strategy for dm to empower and develop its people. The dm approach is also visible in the way it describes apprentices. According to Bauer et al. (2004) the human-centric learning orientation is reflected by calling apprentices *Lernlinge* rather than *Lehrlinge*. Moreover, this title refers to the principle of equality and not to a superior status of mentors vis-à-vis apprentices. In order for one to develop one's individuality, dm offers various platforms such as theater classes where people are able to consciously experience themselves and develop self-awareness (dm drogerie markt 2016d). The approach described also supports creation of an agile organization. Employees at dm are encouraged to use their individual capabilities to respond quickly to customer needs and changes rather than relying on supervisors (Pullig 2016).

Cultivating individual contributions in order to encourage innovation and creativity has been driven by the underlying motivation to enable members to think of the organization as a living system rather than as a rational structure. Moreover, the management of dm inherently came to the conclusion that taking care of people first will make numbers take care of themselves (Pfläging and Selders 2012).

4.4.3 Decentralized Organization

Werner criticized hierarchical approaches for producing ineffective decisions which are not lived by the entire organization (Pullig 2016). Werner advocates a conscious focus on process thinking rather than promoting structural thinking.

Harsch (2016) adds that the culture of self-reliability and freedom supports this structure. Every member of the organization is encouraged to reflect on decisions in a conscious manner instead of relying on standardized guidelines. The company operates in an industry of rapid development and change (Bauer et al. 79). More than 70% of the product assortment gets listed new every year. In order to combine the abilities to cope with a complex business environment while fostering its anthroposophical approach, dm established a decentralized organizational structure (Bauer et al. 2004). Creating an organization based on collaboration and trust changes the perspective from centralized control to decentralized empowerment. This also translated into a different description of roles at dm. For example, *leaders* were renamed *persons responsible* to transform the role from a directive to a counseling character (Pfläging and Selders 2012). Moreover, the decentralization of decisions has been leading to a higher level of employee empowerment combined with an offer by managers to support decisions with personal knowledge and guidance. The company views its headquarters not as a central institution of decision but as a supplier for its stores in terms of knowledge and logistics. This structure transforms a traditional understanding to a service-oriented institution on an equal level with regard to its stores. Hence, the customer-supplier logic is a distinct characteristic for dm and is valid for internal and external interactions. Every store represents an individual unit which is led by a store manager with a high degree of flexibility. For example, store managers at dm are able to decide on additional items beyond the basic product assortment in cooperation with assortment managers to best fit local customer needs. In addition, store managers are not generally limited by regional managers regarding store development. Developing a dm store is not primarily dependent on central specifications rather than on the budget they can develop for the store by themselves. In case of a potential hiring decision, the store manager adds the new employee with all related investments to the store's predicted budget and discusses the needed changes with the regional manager. Afterwards, the personnel marketing department offers the respective materials and documents to post the job offer. This is another example which shows the service relationship between central functions and individual stores (Pullig 2016). Overall, hierarchies have been effectively replaced by individual autonomy, horizontal communication, and cooperation. Werner also emphasized the importance of constantly questioning the status quo in order to mobilize employees and create innovative solutions (Fisbeck-Groh 2004).

4.4.4 Higher Purpose

For dm, the human individual represents the center of all its actions (dm drogerie markt 2016e). With regard to this, Peter Harsch emphasizes that dm views human beings as the end and not the means of business activities. The products and services only represent a means in order to promote human flourishing. The primary goal is to create the business as a social organism which gives its internal and external stakeholders the ability to support this aim. The higher purpose of dm is to create *valuable contributions for society*. The internal dimension of this aspect is reflected by the *creation of a platform for developing and shaping people's lives*. In order to create valuable contributions for society, dm supports various initiatives such as kindergartens, educational programs, sustainability initiatives and other social institutions and projects. In addition, every store can decide on individual projects in order to meet local needs.

Peter Harsch emphasizes that it is very important for dm to contribute its part to support society as a whole. This perspective is also linked to the establishment of consciousness among employees regarding their role in society (Pullig 2016). Creating consciousness regarding shared goals and vision on a store level is identified as a continued field of action. This field supports the development of a systemic view for future opportunities on a local and global scale (Pfläging and Selders 2012).

4.4.5 Stakeholder Integration

Werner (2013) states that individuality is a core element of the humanistic perspective at dm with regard to all stakeholders, especially customers and employees. This approach shows the clear orientation to trust, which is perceived as a necessary requirement (Bauer et al. 2004). Establishing a sense of systemic understanding is needed in order to integrate stakeholder needs. This understanding does not aim at creating formal learning procedures, but to support a holistic way of perceiving interconnected elements of systems. Another guiding principle is *Mein Kollege, mein bester Kunde—mein Kollege, mein bester Lieferant [My colleague, my best customer—my colleague, my best supplier]*. This perspective highlights the importance of internal collaboration. Moreover, dm focuses on a high quality of customer service, for customers are perceived as the main beneficiary among its stakeholders (Pullig 2016). With reference to this view, Harsch (2016) highlights the guiding principle of dm to maximize value for its main recipients, the customers and employees rather than trying to accumulate profits. With regard to this, Peter Harsch emphasizes the active involvement of employees in decisions as it fosters a culture of freedom and self-management which is necessary to meet customer needs. The company focuses on the development of its stakeholders rather than on pursuing profits (Amann and Khan 2011). Harsch notes that this direction aims at investing in a human future which represents a guiding principle for integrating stakeholder interests towards a unified goal. According to Werner (2013), the accumulation of high profits for the sake of profits would show a defect for dm regarding meaningful investment in employees and customers. Transparency is a core value for dm and has direct implications regarding the integration of stakeholders (dm drogerie markt 2016f). Werner rejects the idea of personnel-related costs as he views employees as the source of organizational flourishing. Therefore, he changes the notion from employee-related costs to employee income. This translates into a planning process which assumes the investments required to sustain the business and project the needed amount of capital in order to satisfy internal and external stakeholders.

Calculating an employee's income is oriented to three aspects, *transparent calculation*, *individual agreement*, and *tenability in the community*. In total, there are five basic income levels which reflect the *complexity of the task*, the *chances and risk potential for the organization's success*, and *task-related requirements* such as degree of self-reliance or necessary expertise. The employee's income is evaluated once a year based on an individual development review. Potential income level changes can result according to the range of tasks and responsibilities. In addition, the income is adapted to the yearly performance of the respective work group. In case of an above-expected performance, employees benefit by receiving an additional payment. If the performance ranges below the expectations, additional payments are canceled but the guaranteed income is not affected (Köberlin 2013).

4.4.6 Conscious Leadership

Götz Werner defines the core of leadership as leading one's consciousness (dm drogerie markt 2016g). The founder emphasizes that the task of leadership is to enable as many employees as possible to apply such a mindset and act with a focus on the whole, rather than focusing on a limited area of responsibility. Hence, expert knowledge is viewed as only one part of personal development while the activation of individual abilities to apply knowledge and act consciously is more important (Pullig 2016). The company fosters a collaborative and dialogical leadership approach which is based on shared responsibilities for managers and employees. Every member is encouraged to use his or her own abilities to make responsible decisions. Thus, implementing a culture of individual opportunities and freedom is fostered at dm in contrast to traditional approaches which focus on limits and regulations (Harsch 2016). At dm, group discussions are an essential part of the communication to share empowerment. The learning approach at dm is targeted towards a rather informal and spontaneous learning process because the company aims at a continuous learning-on-the-job approach to enable its employees to benefit from their practical experience (Stiel and Brater 2004). The company does not perceive individuality and collective orientation as contradictory, for individuals need to contribute with their knowledge in order to strengthen the advancement of collective goals. Hence, applying a conscious mindset regarding the impacts of a decision is a guiding principle at dm. In order to establish such a mutually beneficial perspective and entrepreneurial thinking, dm avoids hierarchical leadership models and encourages the use of *recommendations* rather than *instructions* in its concept of dialogical leadership (Pullig 2016).

4.4.7 Conscious Culture and Management

Peter Harsch underlines that fostering a dialogical and collaborative culture is predominantly driven by developing consciousness among every member of the organization. The dialogical culture is present in the company's daily operations and is supported by various seminars and platforms to sustain it. In addition, new apprentices are able to experience the dm culture by participating in theater projects like *Abenteuer Kultur*, which has existed since the year 2000 (dm drogerie markt 2016h). Creating a high-trust culture is a prerequisite for Werner to foster consciousness among individuals in the work context. Creating individual freedom in decisions is another factor which has to be in place to foster innovation and personal initiative as part of a dialogical culture (Pullig 2016). Bauer et al. (2004) add that openness and willingness for personal change and development is considered at a high level at dm because it supports the culture in itself.

With regard to this, Harsch notes that it also requires a high level of awareness to identify tendencies which do not reflect a dialogical culture at dm to avoid a lack in consistency.

4.4.8 Company Performance

Werner (2013) states that dm has never experienced an accumulated loss by the end of a fiscal year. However, he views the accumulation of profits without purposeful utilization as a source of conservative leadership and mental laziness. Werner justifies this perspective with the disappearance of constant learning processes which are a result of a company direction that favors short-term profits rather than the usefulness of its actions. The company has been operated as a limited liability company of which Werner held a stake of 50%. In order to avoid potential inheritance conflicts and further risks associated with this type of ownership, Werner established the dm-Werner foundation. This foundation guarantees that dm continuously aims at creating surpluses only to the extent required to stabilize the company's operations. A potential surplus gets reinvested in the company and additional proceeds are used for philanthropic measures. Moreover, this change leads to the omission of private ownership of Werner's stake. Overall, Werner considers lifetime achievements more successful if they are designed to be successfully led by others. This clearly demonstrates a perspective of depersonalized achievement in order to sustain the founding principle and purpose of dm as an organization (Werner 2013). In terms of its revenue development, dm has been able to achieve a constant increase in the last three years, as outlined in Fig. 4.17.

With reference to the revenue development, the dm Holding has been able to achieve a Compounded Annual Growth Rate of 6% and overall growth of 26.24% since the fiscal year 2012–2013. Pullig (2016) notes that the combination of comparably low prices for basic products combined with a high service orientation for customers has been a major driver for the company's success.

In addition, expansion of the company is driven internally and not through acquisition of other companies. Besides the increase in revenue, dm was also able to increase the number of stores both in Germany and other European countries. The development from 2013 to 2016 is shown in Fig. 4.18.

The number of employees increased as clarified in Fig. 4.19 during the same time.

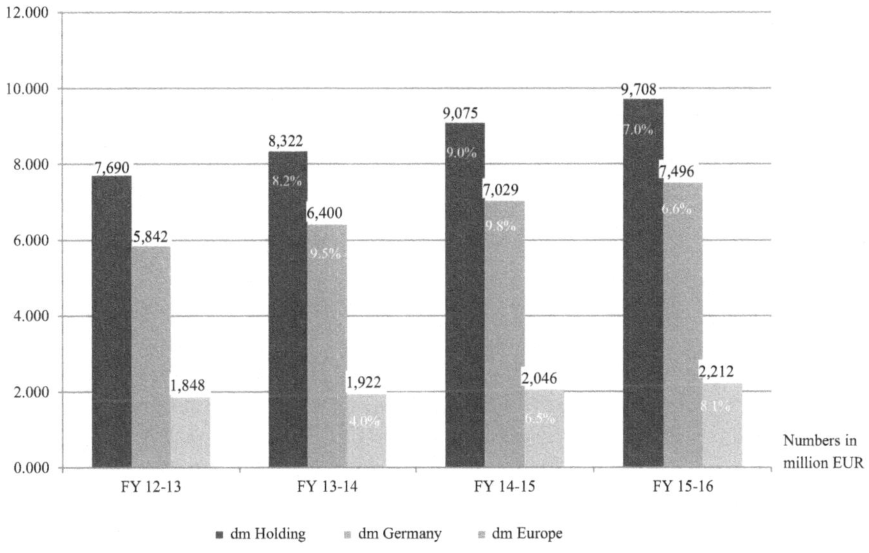

Fig. 4.17 Revenue development and yearly percental change of dm (dm drogerie markt 2016b, 2017b)

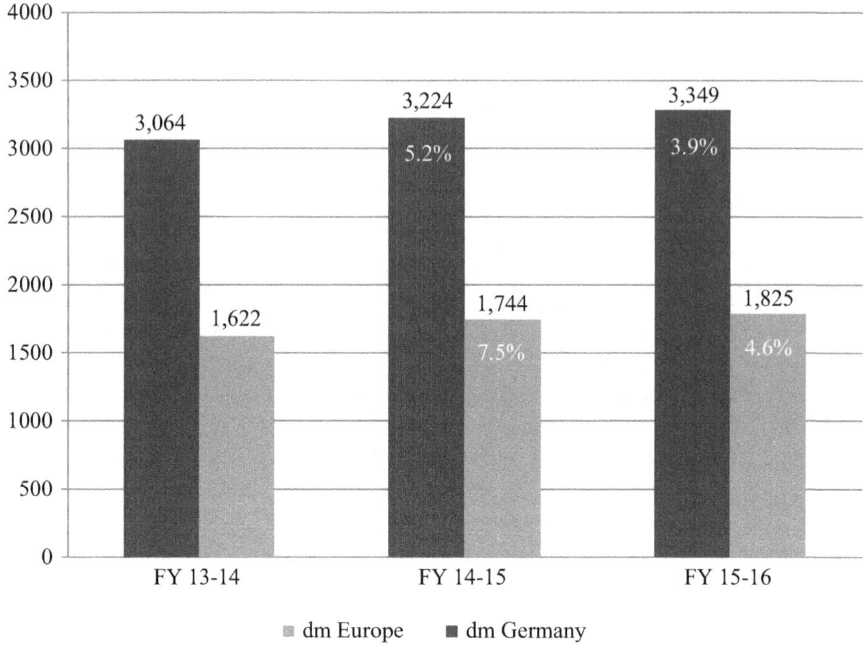

Fig. 4.18 Store expansion and yearly percental change of dm in Germany and Europe (dm drogerie markt 2016b, 2017b)

Statista (2016) has combined research data from 2013 to 2015 and concluded that dm is also the most popular drug store in Germany and has constantly increased in terms of popularity in this time frame. The *Kundenmonitor* nominated dm as the most popular drug store and a segment leader in the categories of *overall satisfaction*, *recommended choice*, *competitive advantages*, and *cost-benefit ratio* (dm drogerie markt 2016i).

With regard to productivity per square meter, dm also leads in its industry with 7000 €/m^2 followed by Rossmann with 5010 €/m^2, and Müller with 4410 €/m^2 (Hahn Gruppe 2013). In addition, Focus (2016) conducted a survey in cooperation with XING and nominated dm as the best employer in Germany in 2016. The results are based on the rating of 70,000 respondents from different levels and functions. In addition to significantly positive financial metrics, dm has been nominated as the most meaningful brand in Germany in 2017. The Havas Group conducted a global survey that was based on more than 300,000 interviews and analyzed more than 1500 companies in 33 countries (Havas Group 2017). The analysis was led by the impact of brands on the well-being of customers on 12 dimensions, which are split into personal benefits, collective benefits and functional benefits. In addition to these aspects, the company content produced has been analyzed regarding its meaningfulness and importance for customers.

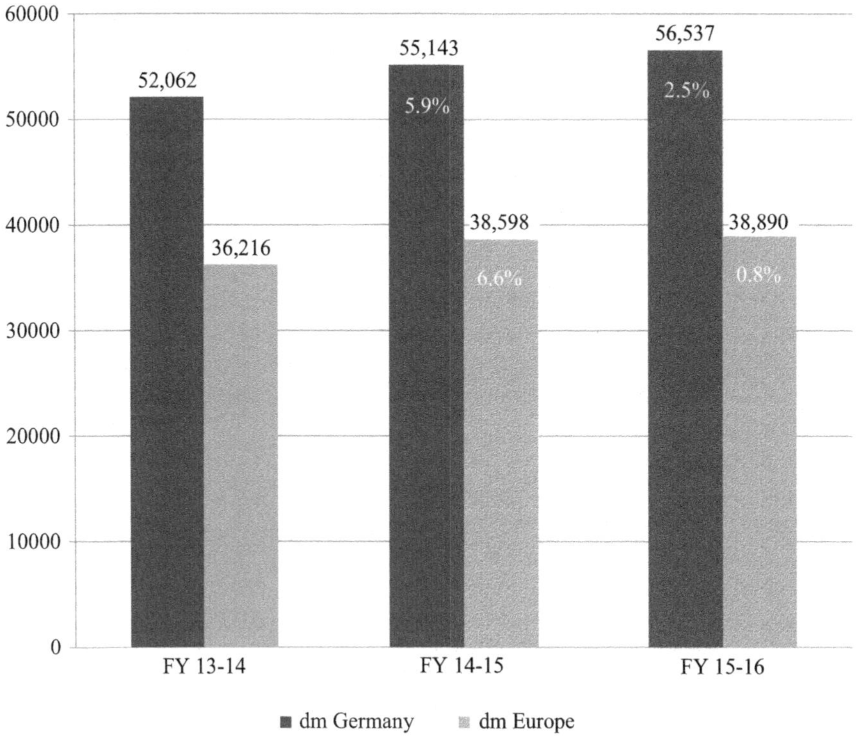

Fig. 4.19 Number of employees at dm in Germany and Europe (dm drogerie markt 2016b, 2017b)

4.4.9 Conclusion

Based on Kofman (2006), dm's anthroposophical philosophy is a clear indicator for the human-centric aspect with regard to a core criterion of conscious business. This orientation is also expressed by dm's slogan *hier bin ich Mensch, hier kauf ich ein* as well as the inherent priority of the company to view individuals as the end and not as a means of its business activities. In addition, the company focuses on the implementation of decentralized structures in order to democratize decision-making processes and establish a high level of agility and self-management. Laloux (2014) recognizes these aspects as core characteristics of conscious businesses. The higher purpose of dm relates to an *altruistic service* for both internal and external stakeholders. Creating valuable contributions to society and a platform to develop and shape people's lives emphasizes dm's orientation to create fulfillment. Mourkogiannis et al. (2008) and Mackey and Sisodia (2014) mention this dimension as a valid orientation of conscious businesses. In terms of integrating stakeholders, dm fosters a systemic view on fulfilling the interests of internal and external stakeholders.

Moreover, dm puts an emphasis on supporting individuality and judges its actions by the creation of value for all, rather than the accumulation of profits. Sisodia et al. (2014) and Freeman et al. (2007) see the rejection of a trade-off mindset regarding stakeholders as an essential aspect of conscious businesses which applies in the case of dm. The leadership approach of dm is based on a dialogic mindset and an acknowledgement of equality among managers and employees. In addition, dm fosters development of individual leadership competences and views leaders as servants in the process of supervising employees in achieving conscious self-leadership. Mackey and Sisodia (2014), Secretan (2001), and Kofman (2006) view these factors as essential for leadership in conscious organizations. The culture of dm is also based on dialogic partnerships and collaboration with a high degree of individual freedom and trust. This contrasts with traditional organizations which usually rely on the establishment of limits and regulations in terms of culture and management.

The decentralized structure of dm can be perceived as an additional factor that favors a conscious management approach, since it leads to a low level of central regulation and benefits from individual strength of its stores and their local orientation. McGrath (2013) and Kofman (2006) regard this structure as a prerequisite and highlight the importance of individual freedom and empowerment. Moreover, the TACTILE classification by Mackey and Sisodia (2014) can be seen as an existent and inherent aspect of dm's culture and management approach. Besides the core aspects of the research framework, dm also demonstrated a consistent above-average performance in its industry. Revenue development, number of stores, number of employees, and further figures indicate the superior performance of the company in its market. In addition to that, dm has been nominated the most popular drug store in Germany and the best employer in Germany according to a survey conducted by Focus (2016) in cooperation with XING.

Superior performance is indicated by Sisodia et al. (2014) as a characteristic of conscious business which can be considered as existent in the case of dm. In addition, dm has been recognized as the most meaningful brand in Germany according to the Havas Group (2017). Overall, dm has a majority of core characteristics and can thus be seen as a representative of the conscious business principles in Germany.

References

Aburdene P (2005) Megatrends 2010: the rise of conscious capitalism. Hampton Roads, Charlottesville, VA

Amann W, Khan S (2011) Dialogue-based leadership style as part of humanistic organizational cultures: the case of dm in Germany. In: Kimakowitz E, Pirson M, Spitzeck H, Dierksmeier C, Amann W (eds) Humanistic management in practice. Palgrave Macmillan, New York, pp 92–102

Assmann J (2016) Start Green Gründerinterview mit Polarstern. https://start-green.net/good-practice/wir-wollen-dazu-beitragen-gesellschaftliche-probleme-zu-loesen/. Accessed 10 July 2016

Backhaus-Maul H, Biedermann C, Nährlich S, Polterauer J (2008) Corporate Citizenship in Deutschland: Bilanz und Perspektiven. VS, Wiesbaden

Bauer HG, Brater M, Büchele U, Dahlem H, Maurus A, Munz C (2004) Lernen im Arbeitsalltag: Wie sich informelle Lernprozesse organisieren lassen. Bertelsmann, Bielefeld

Bertelsmann Stiftung (2016) Social Impact Investment in Deutschland 2016: Kann das Momentum zum Aufbruch genutzt werden? https://www.bertelsmann-stiftung.de/fileadmin/files/BSt/Publikationen/GrauePublikationen/Studie_ZZ_Social-Impact-Investment-in-Deutschland-2016_2016.pdf. Accessed 20 July 2016

BMWi (2017) Gesetz zur Förderung von Mieterstrom und zur Änderung weiterer Vorschriften des Erneuerbare-Energien-Gesetzes Gesetzentwurf der Bundesregierung. http://www.bmwi.de/Redaktion/DE/Downloads/E/entwurf-mieterstrom.pdf?__blob=publicationFile&v=6. Accessed 20 Apr 2017

Deutsches Institut für Service-Qualität (2015) Kundenbefragung Bank des Jahres. http://disq.de/2015/20151118-Bank-des-Jahres.html. Accessed 22 July 2016

dm drogerie markt (2016a) dm-drogerie markt im Zeitverlauf. https://www.dm.de/unternehmen/ueber-uns/historie/. Accessed 12 July 2016

dm drogerie markt (2016b) Unternehmenszahlen. https://www.dm.de/unternehmen/ueber-uns/zahlen-und-fakten/. Accessed 14 July 2016

dm drogerie markt (2016c) Nachhaltigkeit in der Ausbildung. https://www.dm.de/engagement/menschen-bei-dm/-0-c537420.html. Accessed 15 July 2016

dm drogerie markt (2016d) Was die Ausbildung bei dm besonders macht. https://www.dm.de/unternehmen/arbeiten-und-lernen/ausbildung-und-studium/ausbildungskonzept/. Accessed 18 July 2016

dm drogerie markt (2016e) Unternehmen. https://www.dm.de/unternehmen/. Accessed 17 July 2016

dm drogerie markt (2016f) Grundsätze. https://www.dm.de/unternehmen/ueber-uns/grundsaetze/. Accessed 18 July 2016

dm drogerie markt (2016g) dm im Dialog gestalten. https://www.dm.de/engagement/menschen-bei-dm/dm-im-dialog-c535018.html. Accessed 19 July 2016

dm drogerie markt (2016h) Jahrespressekonferenz 2015. https://www.dm.de/unternehmen/medien-und-presse/pressekonferenzen/jahrespressekonferenz-2015/pk-2015-rede-harsch-c736858.html. Accessed 18 July 2016

dm drogerie markt (2016i) dm beliebtestes Unternehmen im Lebensmitteleinzelhandel. https://www.dm.de/unternehmen/medien-und-presse/pressemitteilungen/unternehmen/kundenmonitor-2016-dm-beliebtestes-unternehmen-c878730.html. Accessed 18 Sept 2016

dm drogerie markt (2017a) Logo. https://www.dm.de/unternehmen/medien-und-presse/pressebilder/unternehmen/. Accessed 20 Jan 2017

dm drogerie markt (2017b) Zahlen und Fakten. https://www.dm.de/unternehmen/ueber-uns/zahlen-und-fakten/. Accessed 10 Mar 2017

dm drogerie markt (2017c) dm Online-Verkauf von dm in China erfolgreich gestartet. https://cdn02.dm-static.com/file/de_mcr/c981862/original/1488990763000/pm-dm-china.pdf. Accessed 23 Apr 2017

Dohmen C (2011) Good bank: Das Modell der GLS Bank. Orange, Freiburg

eKomi (2016) Bewertung für Polarstern. https://www.ekomi.de/bewertungen-polarstern.html. Accessed 20 July 2016

Eurosif (2016) European SRI Study 2016. http://www.eurosif.org/wp-content/uploads/2016/11/SRI-study-2016-HR.pdf. Accessed 12 Mar 2017

Fisbeck-Groh SN (2004) Marketing-Entscheidungen im Einzelhandel: Zentralisation versus Dezentralisation. GWV, Wiesbaden

Fitzner M (2009) Markeninvolvement durch Moral: Das Fallbeispiel Vaude. In: Sonnenburg S (ed) Swarm Branding 2.0: Markenführung im Zeitalter von Web 2.0. VS, Wiesbaden, pp 111–124

Focus (2016) FOCUS zeichnet Deutschlands beste Arbeitgeber 2016 aus. http://www.focus.de/finanzen/karriere/focus-top-nationaler-arbeitgeber-focus-zeichnet-deutschlands-beste-arbeitgeber-2016-aus_id_5247823.html. Accessed 8 July 2016
Freeman ER, Harrison JS, Wicks AC (2007) Managing for stakeholders: survival, reputation, and success. Yale University Press, New Haven, CT
Friedman AL, Miles S (2006) Stakeholders: theory and practice. Oxford University Press, Oxford
Fueglistaller U, Müller C, Volery T (2004) Entrepreneurship: Modelle—Umsetzung—Perspektiven: Mit Fallbeispielen aus Deutschland, Österreich und der Schweiz. Gabler, Wiesbaden
George B (2014) Getting capitalism back on track. In: Mackey J, Sisodia R (eds) Conscious capitalism: liberating the heroic spirit of business. Harvard Business Review, Boston, MA, pp ix–xiv
Global Alliance for Banking on Values (2016) About us. http://www.gabv.org/about-us. Accessed 20 July 2016
GLS Bank (2013) Nachhaltigkeitsbericht 2013. https://www.gls.de/download/?file=4467&cHash=176cbf9390b51a41a8ac5609d9cc6579. Accessed 18 July 2016
GLS Bank (2014a) Leitbild. https://www.gls.de/media/pdf/Broschueren/GLS_Bank/gls_leitbild.pdf. Accessed 18 July 2016
GLS Bank (2014b) Bilanzpressekonferenz der GLS Bank. https://www.gls.de/download/?file=4799&cHash=be7276d1b98d22b4f3f443a57b9b9837. Accessed 20 July 2016
GLS Bank (2015a) GLS Bank für vorbildliche Geschäftspolitik ausgezeichnet.https://www.gls.de/privatkunden/ueber-die-gls-bank/presse/pressemitteilungen/gls-bank-fuer-vorbildliche-geschaeftspolitik-ausgezeichnet/. Accessed 18 July 2016
GLS Bank (2015b) Nachhaltigkeitsbericht 2015. https://www.gls.de/download/?file=8037&cHash=44be6616420a740ba83d795fe574c7da. Accessed 18 July 2016
GLS Bank (2016a) GLS Geschichte: Erfahren Sie mehr über Ihre Bank. https://www.gls.de/privatkunden/ueber-die-gls-bank/geschichte/. Accessed 12 July 2016
GLS Bank (2016b) Nachhaltigkeit und Werte. https://www.gls.de/privatkunden/ueber-die-gls-bank/nachhaltigkeit-werte-und-auszeichnungen/. Accessed 19 July 2016
GLS Bank (2016c) Der Bankspiegel—unsere Kundenzeitschrift. https://www.gls.de/privatkunden/aktuelles/bankspiegel/. Accessed 19 July 2016
GLS Bank (2016d) 2015: Auf Zukunftskurs mit erweiterter Geschäftsleitung. https://www.gls.de/privatkunden/ueber-die-gls-bank/transparenz/zahlen-und-fakten/. Accessed 19 July 2016
GLS Bank (2016e) Wie viele Mitglieder und Kunden hat die GLS Bank?. https://www.gls.de/privatkunden/service-kontakt/suchen-und-finden/fragen-antworten/ueber-die-gls-bank-und-ihre-arbeit/gls-bank/wie-viele-mitglieder-und-kunden-hat-die-gls-bank/. Accessed 19 July 2016
GLS Bank (2016f) Für eine lebenswerte Zukunft. https://www.gls.de/privatkunden/kundenportraets. Accessed 19 July 2016
GLS Bank (2016g) GLS Anlage- und Finanzierungsgrundsätze. https://www.gls.de/privatkunden/ueber-die-gls-bank/arbeitsweisen/anlage-und-finanzierungsgrundsaetze/. Accessed 20 July 2016
GLS Bank (2016h) Ethibel-Ranking 2005: Beste Bewertung in Deutschland. https://www.gls.de/privatkunden/ueber-die-gls-bank/nachhaltigkeit-werte-und-auszeichnungen/auszeichnungen/ethibel-ranking-2005-beste-bewertung-in-deutschland/. Accessed 20 July 2016
GLS Bank (2017a) Pressematerial. https://www.gls.de/privatkunden/ueber-die-gls-bank/presse/pressematerial/. Accessed 11 Apr 2017
GLS Bank (2017b) Zahlen & Fakten. https://www.gls.de/privatkunden/ueber-die-gls-bank/transparenz/zahlen-und-fakten/. Accessed 11 Mar 2017
Hahn Gruppe (2013) Retail real estate report Germany. http://www.cbre.de/de_de/research/retail/retail_content/CBRE_Hahn_Retail_Report_Deutschland_2013-2014_deutsch.pdf. Accessed 4 July 2016
Harsch E (2016) Personal interview

Havas Group (2017) Meaningful brands. http://www.havasmedia.de/organic-marketing/meaningful-brands/. Accessed 15 May 2017
Hay Group (2013) Fluktuation weltweit 2013 und 2018. http://www.haygroup.com/downloads/de/Tabelle%20-%20Fluktuation%20weltweit%202013%20und%202018.pdf. Accessed 18 July 2016
Henle F (2016) Personal interview
International Association of Investors in the Social Economy (2016) About us. http://www.inaise.org/?q=en/node/15. Accessed 21 July 2016
Jensen A (2011) Wir steigern das Bruttosozialglück: Von Menschen, die anders wirtschaften und besser leben. Herder, Freiburg
Jorberg T (2016a) Was die GLS Bank anders macht. In: Wendt K (ed) CSR und Investment Banking: Zwischen Krise und Positive Impact. Springer, Berlin, pp 283–295
Jorberg T (2016b) GLS Bank: Chef Thomas Jorberg über Crowdfunding und Sondergebühren. Wirtschaftswoche 51:84
Kampffmeyer N, Knopf J (2011) Die Bedeutung von Unternehmenskultur und Netzwerken für Nachhaltigkeits innovationen am Beispiel der GLS Bank. In: Knopf J, Quitzow R, Hoffmann E, Rotter M (eds) Nachhaltigkeitsstrategien in Politik und Wirtschaft: Treiber für Innovation und Kooperationen. oekom, München, pp 149–170
Kerler R (2011a) Ändere dich selbst: So änderst du die Gesellschaft: Im Gespräch mit dem Bankmitbegründer und Soziologen Rolf Kerler. In: Dohmen C (ed) Good Bank: Das Modell der GLS Bank. Orange, Freiburg, pp 36–49
Kerler R (2011b) Eine Bank für den Menschen: Von den Anfängen und Impulsen der GLS Treuhand und GLS Bank. Verlag am Goetheanum, Hamburg
Köberlin T (2013) Systemische Unterstützung der Mitarbeitereinsatzplanung bei dm drogerie markt. In: Bornewasser M, Zülch G (eds) Arbeitszeit—Zeitarbeit. Wiesbaden, Gabler, pp 333–343
Kofman F (2006) Conscious business: how to build value through values. Sounds True, Boulder
KPMG (2015) European Responsible Investing Fund Survey 2015. http://www.alfi.lu/sites/alfi.lu/files/European-Responsible-Investing-Fund-Survey-2015_1.pdf. Accessed 20 June 2016
Kununu (2016) GLS Bank. https://www.kununu.com/de/gls-bank. Accessed 21 July 2016
Laloux F (2014) Reinventing organizations: a guide to creating organizations inspired by the next stage of human consciousness. Nelson Parker, Brussels
Lerner RM, Easterbrooks MA, Mistry J (2003) Handbook of psychology, volume 6, Developmental psychology. Wiley, Hoboken
Mackey J, Sisodia R (2014) Conscious capitalism: liberating the heroic spirit of business. Harvard Business School Publishing, Boston, MA
McGrath A (2013) Holocracy: creating conscious organizations. http://wonderworksconsulting.com/wp-content/uploads/2013/10/Creating_Conscious_Org.pdf. Accessed 18 June 2016
Mourkogiannis N, Vogelsang G, Unger S (2008) Der Auftrag: Was großartige Unternehmen antreibt. Wiley-VCH, Weinheim
Pfläging N, Selders J (2012) Beyond Budgeting: Dezentralisierte Führung und flexible Steuerungsprozesse umsetzen am Beispiel der Wertbildungsrechnung bei dm drogerie markt. In: Dillerup R, Stoi R (eds) Fallstudien zur Unternehmensführung. Vahlen, München, pp 397–414
Polarstern (2015) Gemeinwohlbericht. https://www.polarstern-energie.de/fileadmin/files/blog/160225/gemeinwohl-oekonomie-bilanz-polarstern.pdf. Accessed 9 July 2016
Polarstern (2016a) 5 Jahre Polarstern! Ein Gespräch mit den Gründern. https://www.polarstern-energie.de/magazin/5-jahre-polarstern-ein-gespraech-mit-den-gruendern/. Accessed 5 July 2016
Polarstern (2016b) Kurzprofil. https://www.polarstern-energie.de/presse/informationen/#kurzprofil. Accessed 8 July 2016

Polarstern (2016c) Die Entscheidung für einen Ökogas-Anbieter ist nicht so einfach—VAUDE. https://www.polarstern-energie.de/presse/mitteilung/die-entscheidung-fuer-einen-oekogas-anbieter-ist-nicht-so-einfach-vaude/. Accessed 9 July 2016

Polarstern (2016d) Polarstern veröffentlicht als erster Energieversorger eine Gemeinwohlbilanz. https://www.polarstern-energie.de/presse/mitteilung/polarstern-veroeffentlicht-als-erster-energieversorger-eine-gemeinwohlbilanz/. Accessed 9 July 2016

Polarstern (2016e) Magazin. https://www.polarstern-energie.de/magazin/. Accessed 8 July 2016

Polarstern (2016f) Weltweit. https://www.polarstern-energie.de/weltweit/. Accessed 7 July 2016

Polarstern (2016g) Die Top 100 der Sinn-Ökonomie. https://www.polarstern-energie.de/magazin/die-top-100-der-sinn-oekonomie/. Accessed 12 July 2016

Polarstern (2016h) Gemeinwohl-Ökonomie: Polarstern ist der erste Energieversorger mit Gemeinwohl-Bilanz. https://www.polarstern-energie.de/magazin/gemeinwohl-oekonomie-polarstern-ist-der-erste-energieversorger-mit-gemeinwohl-bilanz/. Accessed 8 July 2016

Polarstern (2016i) Kampf dem Isar-Müllberg. https://www.polarstern-energie.de/presse/mitteilung/kampf-dem-isar-muellberg/. Accessed 12 July 2016

Polarstern (2017a) Logo. https://www.polarstern-energie.de/presse/informationen/#logos. Accessed 20 Apr 2017

Polarstern (2017b) Wirklich Ökostrom für Wärmepumpen. https://www.polarstern-energie.de/panasonic/. Accessed 15 Apr 2017

Pullig K (2016) Innovative Unternehmenskulturen: Zwölf Fallstudien zeitgemäßer Sozialordnungen. Springer, Wiesbaden

Secretan L (2001) The conscious leader. Ind Week 250(2):19–20

Sisodia R, Sheth J, Wolfe D (2014) Firms of endearment: how world-class companies profit from passion and purpose. Pearson, New York

Statista (2016) Ranking der beliebtesten Drogeriemärkte in Deutschland in den Jahren 2013 bis 2015. http://de.statista.com/statistik/daten/studie/171495/umfrage/in-den-letzten-6-monaten-besuchte-drogeriemaerkte/. Accessed 14 July 2016

Stein L (2010) Managementpraktiken unternehmerischer Nachhaltigkeit: Wie Unternehmen ihren Beitrag zu einer lebenswerten Welt gestalten. Carl-Auer, Heidelberg

Stiel M, Brater M (2004) Die lernenden Filialen. Eine Fallstudie zur Mitarbeiter- und Organisationsentwicklung. http://www.brainguide.de/upload/publication/18/kcos/46866f6a2f1d2f067b42e40d6a42edcc_1311535223.pdf. Accessed 8 July 2016

UBS (2015) Adding value(s) to investing: sustainable investing. https://www.ubs.com/content/dam/WealthManagementAmericas/documents/adding-values-to-investing-2015-03-sustain able-investing.pdf. Accessed 21 July 2016

VAUDE (2014) GWÖ-Bericht. http://nachhaltigkeitsbericht.vaude.com/gri-wAssets/pdf/de/2014_VAUDE_GWOe_Testat.pdf. Accessed 8 July 2016

VAUDE (2015) PFC in outdoor products—balancing performance and ecology. https://vaude.zendesk.com/hc/en-us/articles/203898278--05-02-2015-PFC-in-outdoor-products-balancing-performance-and-ecology. Accessed 22 July 2016

VAUDE (2016a) VAUDE facts and numbers. http://www.vaude.com/en-GB/VAUDE/Media/Facts-Numbers/. Accessed 9 July 2016

VAUDE (2016b) VAUDE Green Shape. http://nachhaltigkeitsbericht.vaude.com/gri/produkte/greenshape-konzept.php. Accessed 12 July 2016

VAUDE (2016c) The VAUDE Ecosystem. http://csr-report.vaude.com/gri-en/vaude/vaude-eco system.php. Accessed 18 July 2016

VAUDE (2016d) Nachhaltiges Wirtschaften lohnt sich. http://nachhaltigkeitsbericht.vaude.com/gri/vaude/nachhaltiges-wirtschaften.php. Accessed 16 July 2016

VAUDE (2016e) Das sind unsere Stakeholder. http://nachhaltigkeitsbericht.vaude.com/gri/vaude/Stakeholder-im-Detail.php. Accessed 17 July 2016

VAUDE (2016f) Beruf und Privatleben im Einklang. http://nachhaltigkeitsbericht.vaude.com/gri/menschen/beruf-und-privatleben.php. Accessed 17 July 2016

VAUDE (2016g) Wirtschaften zum Wohle aller. http://nachhaltigkeitsbericht.vaude.com/gri/csr-standards/gemeinwohloekonomie.php. Accessed 20 July 2016
VAUDE (2016h) GWÖ-Bilanz. http://nachhaltigkeitsbericht.vaude.com/gri-wAssets/pdf/de/2014_VAUDE_GWOe_Testat.pdf. Accessed 14 July 2016
VAUDE (2016i) Vertrauenskultur. http://nachhaltigkeitsbericht.vaude.com/gri/menschen/vertrauenskultur.php. Accessed 18 July 2016
VAUDE (2016j) Kundenzufriedenheit als höchstes Ziel. http://nachhaltigkeitsbericht.vaude.com/gri/vaude/kundenzufriedenheit.php. Accessed 19 July 2016
VAUDE (2016k) Zufriedene Mitarbeiter sind engagiert und motiviert. http://nachhaltigkeitsbericht.vaude.com/gri/menschen/mitarbeiterzufriedenheit.php. Accessed 19 July 2016
VAUDE (2016l) Aus- und Weiterbildung für alle Mitarbeiter. http://nachhaltigkeitsbericht.vaude.com/gri/menschen/aus-und-weiterbildung.php. 19 July 2016
VAUDE (2017a) Logo. http://mam.vaude.com/opas/view_pin_cart.jsf;jsessionid=5B861BBC047D1EEE28261CA1B41F94C3. Accessed 20 Apr 2017
VAUDE (2017b) Green shape. http://mam.vaude.com/opas/view_pin_cart.jsf. Accessed 20 Feb 2017
VAUDE (2017c) VAUDE wins A.T. Kearney Family Award—most family-friendly company. https://www.vaude.com/en-GB/Company/Media/VAUDE-wins-A.T.-Kearney-Family-Award. Accessed 10 Apr 2017
VAUDE (2017d) VAUDE receives Soul@Work Award—excellent employee management. https://www.vaude.com/en-GB/Company/Media/VAUDE-receives-SoulatWork-Award. Accessed 14 Apr 2017
VAUDE (2017e) VAUDE—strong in Europe. http://csr-report.vaude.com/gri-en/vaude/our-brand.php. Accessed 11 Apr 2017
Von Dewitz A (2016) Personal interview
Weber M, Scheck B (2012) Impact Investing in Deutschland Bestandsaufnahme und Handlungsanweisungen zur Weiterentwicklung. http://www.bertelsmannstiftung.de/fileadmin/files/user_upload/Impact_Investing_in_Deutschland.pdf. Accessed 14 June 2016
Weibel A, Schafheitle S (2015) Trust in the employer. http://www.ifpm.unisg.ch/~/media/internet/content/dateien/instituteundcenters/ifpm/jahresbericht_2015.pdf. Accessed 19 July 2016
Werner G (2007) Der Ästhet. In: Koch H (ed) Soziale Kapitalisten: Vorbilder für eine gerechtere Wirtschaft. Rotbuch, Berlin, pp 60–71
Werner G (2013) Womit ich nie gerechnet habe. Econ, Berlin
Wuczkowski M (2015) Biodiversität und Unternehmen: Eine Untersuchung nachhaltigkeitsorientierter organisationaler Lernprozesse zur Umsetzung in der betrieblichen Praxis. Metropolis, Marburg
Zajic J (2016) Personal interview

Chapter 5
Discussion, Indications and Conclusion

The findings of the case studies are discussed based on the study conducted and contrasts the results with information from the research framework and interviews conducted. The research hypotheses from Sect. 2.8 are evaluated based on an explorative and qualitative approach as qualitative data cannot be evaluated by a statistical hypotheses test. The first hypothesis assumes that conscious business practices have been contributing to a long-term sustainable growth model and a high degree of organizational performance for conscious businesses in Germany. All interview partners agreed that applying a human and purpose-driven mindset in business is both beneficial for its stakeholders as well as for the business itself. This perspective also correlates with the findings of a study conducted by Harvard Business Review Analytic Services (HBRAS) sponsored by the EY Beacon Institute. Thus, more than 80% of the 474 executives surveyed believe that a strong sense of purpose positively affects an organization's ability to transform and increase customer loyalty (HBRAS 2015). In relation to these results, the 2016 Workforce Purpose Index concluded that employees experience a significantly higher level of satisfaction and commitment when they are working in a purpose-driven environment (LinkedIn and Imperative 2016).

Interview partner 1 states that companies are able to benefit from these principles because they lead to a higher degree of employee performance as a result of increased job satisfaction. The CEO of VAUDE, Antje von Dewitz (2016) perceives it as the basis of long-term success and Erich Harsch is convinced that success is a result of centering business activities on humanity.

Florian Henle (2016) adds that businesses can profit from this perspective by creating high levels of trust among stakeholders, which is essential for long-term success. Janina Zajic highlights that these principles result in a high level of employer attractiveness, an increased agility towards changes in the business environment and more creativity. In addition, VAUDE, dm drogerie markt, and the GLS Bank demonstrated above-average financial performance for multiple years in a row, which can be seen as another indicator of their long-term success. Polarstern was founded 5 years ago and can still be considered as a start-up. As the

© Springer International Publishing AG 2018

N.J. Stahlhofer et al., *Conscious Business in Germany*, CSR, Sustainability, Ethics & Governance, https://doi.org/10.1007/978-3-319-69739-0_5

company represents a social business, it is not oriented towards achieving high profits for investors, but to contributing to the creation of a common good. However, Polarstern has achieved a high level of customer satisfaction which is also a significant characteristic of conscious businesses. The findings of Laloux (2014), Bruch and Ghoshal (2003), and Sisodia et al. (2014) correspond to the arguments described by the interview partners with regard to a company's long-term success. Based on the findings from the case studies, the first hypothesis is regarded as confirmed because all companies analyzed were found to operate on a successful business model on a long-term perspective.

The second hypothesis states that consciousness in business is predominantly driven by a demographic value change and a constant growth and accessibility of information. Zajic notes that an increasing number of people in Germany consciously reflect on their values as a result of a high standard of living according to the satisfaction of basic needs. Von Dewitz is also convinced that a change of values is a primary driver for this concept in Germany. However, Harsch views a change in values not as specifically driven by Millennials but a cross-generational development. Florian Henle applies the same logic as Zajic and hence sees the high standard of living as a driving prerequisite which fostered these principles. Moreover, Henle identifies the dysfunctionality of the current economic system in terms of value distribution and the conscious perception of unethical corporate behavior as a driving force. Interview partner 1 (2016) perceives an increasing demand for fulfilling and enriching work environments by employees as the key driver. With regard to the influence of a generational value change, Matthies (2016) emphasizes the growing number of Millennials in companies as a predominant driver since this generation is strongly purpose-driven. In addition, Peter Matthies highlights the increasing number of global challenges such as climate change, globalization, and increasing pressure at work, as important drivers to fundamentally change the nature of business. A constant growth and accessibility of information are not mentioned by the interview partners as driving factors for this concept. Based on the interview partners' statements, the second hypothesis cannot be clearly considered as confirmed although a change of values has been identified as a key driver. Moreover, all interview partners agree that transparency has become increasingly important for business. Therefore, it could be the subject of further research to clarify the role of an accelerating information society in respect to emergence of conscious businesses.

The third hypothesis relates to the development of conscious business principles in Germany and their relevance in the future. Accordingly, the hypothesis assumes that conscious businesses in Germany have increased in importance and hence will challenge traditional businesses practices towards a new paradigm in the future. Interview partner 1 (2016) is convinced that the importance of conscious organizational cultures will increase because this entails a competitive advantage for businesses.

However, it will take time for most corporations to perceive this aspect as a success factor and act accordingly (Interview partner 1 2016). Peter Matthies notices a growing demand for implementing conscious business principles as companies are increasingly under pressure to cope with global challenges and the value changes in society. The results of the Harvard Business Review study on purpose clearly show a needed shift towards purpose-driven organizations from an executive perspective (HBRAS 2015). These findings are also in line with the conclusion of the 2016 Workforce Purpose Index that shows a clear demand from an employee perspective (LinkedIn and Imperative 2016). Overall, the importance of conscious business principles has been growing worldwide and is likely to continue this development in the future. With reference to this, Janina Zajic confirms a positive development of these principles over the past decades and expects an increasing demand that is also visible by the constant development of the GLS Bank. Florian Henle also believes that these principles will increase in importance in the future due to their systemic benefits. Antje von Dewitz perceives the pursuit of the sustainable concept as the most meaningful way of conducting business. However, she also mentions that the majority of businesses are still not operating their business in such a way. Erich Harsch takes on a similar perspective as he already experienced positive changes in the business environment but also notes that hierarchical organizations are still dominant. Nevertheless, he views it as an ongoing change process and a question of human development that will lead to greater importance of the principles (Harsch 2016). With regard to this Mackey and Sisodia (2014) and Benett et al. (2009) point out that the increasing representation of Millennials and their orientation to pursuing purpose beyond profits is likely to fuel this development. If the principles of conscious business are to be fostered, a variety of framework conditions will have to be embraced actively by companies in the near future. Interview partner 1 (2016) states that acceptance by executives is the key determining factor, whether these principles are pursued internally. With regard to this, Peter Matthies holds a similar opinion and adds that the employees play an essential role after the executive's commitment, since the principles are unlikely to be sustained unless they are accepted actively by the remaining 99% of the organization's members. Antje von Dewitz also mentions that a political and systemic change needs to be fostered which focuses on the creation of a common good and not the maximization of shareholder value. Moreover, von Dewitz and Henle mention a higher level as prerequisites as well as the development and reorientation of economic studies that focus on an ethical way of conducting business. Zajic (2016) adds that it also takes an increasing number of positive examples by companies applying the principles successfully and with an intrinsic motivation. In addition, the B Corporations and the Economy for the Common Good have both increased in terms of popularity and number of certified companies. Based on this development and the increasing awareness of these initiatives, it is likely that this movement will contribute to an emergence of conscious business principles in Germany as well. However, the interview partners have different opinions on the named initiatives. Erich Harsch believes that social interactions cannot be classified by these approaches as interactions develop as a process and cannot be classified by a static structural framework.

Florian Henle, Antje von Dewitz, Zajic (2016), and Peter Matthies perceive the movement as a positive sign towards a higher level of transparency and commitment with reference to the principles of conscious businesses.

Nevertheless, interview partner 1 (2016) points out that certifications without the actual intrinsic implementation of the principles will likely discourage members of a certified organization. Based on the results from the interviews and success of the companies selected for this study, the third hypothesis can be considered as confirmed. With reference to the popularity of the concept, it has to be clarified that the term conscious business was only familiar for interview partner 1, Peter Matthies, and Florian Henle. Nevertheless, all interview partners said that they were familiar with the concepts that relate to conscious business. With regard to this, Matthies mentions that conscious business is well-known among American companies and the term seems to be becoming more popular in Europe as well. German companies usually frame the implementation of conscious business principles as *changes of their corporate culture*, whereas American companies tend to use the term conscious business. Peter Matthies adds that the term itself represents an important anchor in order to foster the principles behind it. However, it is more important to promote and implement the principles rather than focus on the label itself.

5.1 Suggestions for Future Research

This study embodies a non-experimental study design that integrated four companies which differ in size, industry, and ownership. With regard to the respective industries, it would be beneficial to compare the findings regarding the four companies to their competitors. This direction could contrast the impact of conscious business principles in comparison to traditional business concepts. Moreover, other companies from each industry that relate to the fundamental orientation of conscious business could be integrated to gain additional insight regarding the different implementation. For example, the implementation practices of the GLS Bank could be contrasted with various practices of the Ethik Bank. In addition, it could be valuable to analyze the effect of different models of company ownership regarding the freedom to implement principles of conscious business. The companies selected for this study do not include a publicly listed company and therefore are likely to operate on a higher level of independency.

5.2 Summary and Conclusion

The objective of this study was the assessment of the current situation of conscious business principles in German companies. In addition, this book aims at finding evidence that contrasts information regarding the future development of the conscious business concept in Germany. For this reason, the GLS Bank, dm drogerie markt, VAUDE, and Polarstern have been selected to create an insight regarding the implementation of conscious business principles. For the evaluation of these implementations, a research framework has been used in accordance with the conscious business tenets by Mackey and Sisodia (2014). Hence, each company has been explored regarding the elements *higher purpose, stakeholder integration, conscious leadership, conscious culture and management*, as well as its *humanistic orientation*, which, according to Kofman (2006), is an inherent characteristic of conscious businesses. The research was based on interviews with representatives of each business, as well as available information in books, articles, company reports, and other data communicated by the respective companies.

To broaden the view on this research topic, two additional interviews have been conducted on this research topic. Three research hypotheses have been formulated in order to direct the findings towards central aspects that relate to the objectives of the book. With regard to the research hypotheses, two out of three can be considered as confirmed. Based on the findings from the case studies and corresponding results of the theoretical framework, conscious business practices have been contributing to a long-term sustainable growth model and a high degree of organizational performance for conscious businesses in Germany. In addition, it can be assumed that conscious businesses in Germany have increased in importance and hence will challenge traditional business practices towards a new paradigm in the future. However, the second hypothesis cannot be considered as confirmed based on the results from the interviews. Therefore, it cannot be assumed that an evolving consciousness in business is predominantly driven by value change caused by demographic developments, constant growth and accessibility of information. With regard to this hypothesis, the interview partners stated a variety of primary drivers such as general value changes in society, increasing globalization, global challenges such as climate change, unethical company behavior, and an increasing orientation towards purpose in life which has been enabled by a high standard of living in Germany and the related satisfaction of basic needs. Overall, the explorative approach aimed at integrating different perspectives, social and economic drivers, company examples of conscious business practices in the United States, and the inclusion of related concepts such as the B Corporations and the Economy for the Common Good. The qualitative data analysis was conducted according to the principles of Mayring (2010). In addition, quantitative information has been integrated as well to assess the companies' performance and gather additional insight. Based on the findings of this study, the German businesses selected operate on the basis of the principles of conscious business and are able to create a common good by pursuing a purpose beyond profits, integrating stakeholders in a synergetic manner, and pursuing a conscious approach to culture, leadership, and management based on a humanistic orientation. This orientation also translates into multiple areas of corporate performance such as an above-average revenue development and high levels of stakeholder satisfaction.

Overall the principles of conscious business constitute a valuable approach to integrating an economic, ecologic, spiritual, and social dimension in a humanistic and beneficial manner. Moreover, it represents a valuable model for coping with tomorrow's challenges. Peter Drucker said that the best way to predict the future is to create it (Hesselbein 2010).

References

Benett A, Gobhai C, O'Reilly A, Welch G (2009) Good for business: the rise of the conscious corporation. Palgrave Macmillian, New York

Bruch H, Ghoshal S (2003) Unleashing organizational power. http://sloanreview.mit.edu/article/unleashing-organizational-energy/. Accessed 24 Mar 2016

Harsch E (2016) Personal interview

HBRAS (2015) The business case for purpose. http://www.ey.com/Publication/vwLUAssets/ey-the-business-case-for-purpose/$FILE/ey-the-business-case-for-purpose.pdf. Accessed 12 Feb 2017

Henle F (2016) Personal interview

Hesselbein F (2010) Foreword. In: Cohen WA (ed) Drucker on leadership: new lessons from the father of modern management. Jossey-Bass, San Francisco, CA, pp ix–xi

Interview Partner 1 (2016) Personal interview

Kofman F (2006) Conscious business: how to build value through values. Sounds True, Boulder, CO

Laloux F (2014) Reinventing organizations: a guide to creating organizations inspired by the next stage of human consciousness. Nelson Parker, Brussels

LinkedIn and Imperative (2016) 2016 Workforce Purpose Index. Purpose at work. The largest global study on the role of purpose in the workforce. https://cdn.imperative.com/media/public/Global_Purpose_Index_2016.pdf. Accessed 12 Mar 2017

Mackey J, Sisodia R (2014) Conscious capitalism: liberating the heroic spirit of business. Harvard Business School Publishing, Boston, MA

Matthies P (2016) Personal interview

Mayring P (2010) Qualitative Inhaltsanalyse. In: Flick U, von Kardorff E, Steinke I (eds) Qualitative Forschung. Rowohlt, Reinbek, pp 468–475

Sisodia R, Sheth J, Wolfe D (2014) Firms of endearment: how world-class companies profit from passion and purpose. Pearson, New York

Von Dewitz A (2016) Personal interview

Zajic J (2016) Personal interview

Printed by Printforce, the Netherlands